Radiographic Critique

Instructor's Manual

Kathy McQuillen-Martensen, RT(R)
Radiography Instructor, Program in Radiologic Technology
Department of Radiology
The University of Iowa Hospitals and Clinics
Iowa City, Iowa

W.B. SAUNDERS COMPANY
A Division of Harcourt Brace & Company
Philadelphia • London • Toronto • Montreal • Sydney • Tokyo

W.B. SAUNDERS COMPANY
A Division of Harcourt Brace & Company

The Curtis Center
Independence Square West
Philadelphia, Pennsylvania 19106

Radiographic Critique Instructor's Manual   0-7216-4979-3

Copyright © 1996 by W.B. Saunders Company.

All rights reserved. No part of this publication may be reproduced or transmitted in any form or by any means, electronic or mechanical, including photocopy, recording, or any information storage and retrieval system, without permission in writing from the publisher.

Printed in the United States of America

Last digit is the print number: 9   8   7   6   5   4   3   2   1

Preface

This instructor's manual is written in an outline format. It follows the same systematic order that is found in several positioning textbooks and that I use to teach radiographic positioning and critique. Under each section I have described activities that the students in my radiography program complete in the classroom and laboratory settings. These activities have helped the students to effectively understand the information presented in the class lectures. There is also an extensive bank of test questions for each chapter.

I believe that by sharing our innovative methods of teaching positioning and critique with each other, we all will discover some new interactive ways for students to learn this information. If you have any ideas you would like to share with me, I would be happy to listen.

Kathy

Contents

Chapter 1, Guidelines for Radiographic Critique..1

    Final exam test bank..2

Chapter 2, Radiographic Critique of the Chest and Abdomen................................6

    Final exam test bank..8

Chapter 3, Radiographic Critique of the Upper Extremity....................................16

    Final exam test bank..16

Chapter 4, Radiographic Critique of the Shoulder..23

    Final exam test bank..24

Chapter 5, Radiographic Critique of the Lower Extremity....................................31

    Final exam test bank..32

Chapter 6, Radiographic Critique of the Hip and Pelvis......................................41

    Final exam test bank..42

Chapter 7, Radiographic Critique of the Cervical and Thoracic Vertebrae................45

    Final exam test bank..46

Chapter 8, Radiographic Critique of the Lumbar Vertebrae, Sacrum, and Coccyx.....53

    Final exam test bank..54

Chapter 9, Radiographic Critique of the Sternum and Ribs.................................59

    Final exam test bank..60

Chapter 10, Radiographic Critique of the Cranium..62

    Final exam test bank..63

Chapter 1
Guidelines for Radiographic Critique

I. Introduction lecture on radiographic positioning and critique guidelines.

II. Classroom activities.

    A. Workbook study questions.

        The workbook study questions are designed to encourage the student to form mental images of the radiographs and the patient positioning. It is this mental imaging that helps students to understand the information rather than memorizing it. The information is entirely too vast to memorize. Students who are memorizing this information soon begin to have problems retaining the knowledge. They must be taught to form mental images.

    B. Workbook learning activities numbers 1, 2, 3, and 12.

        Suggestion for learning activity number 1: Write all the different radiograph action terms onto separate pieces of paper, then place them in an envelope. Divide the class into two groups. Use the action terms as words to be used in a game of charades. Have fun!

        Suggestion for learning activity number 2: Place radiographs of assorted body parts and positions in a jacket. Instruct the students to hang the radiographs correctly on the view box.

        Suggestion for learning activity number 3:
Accuracy of patient identification plate placement: In a jacket place radiographs that have the ID plate positioned in the best possible location and in the worst location. Instruct the students to evaluate ID placement on the radiographs.

        Patient and facility identification: In a jacket place radiographs that do and do not have all the required facility and patient identification on them. Instruct the students to evaluate the radiographs for proper identification requirements.

        Accuracy of film marking: In a jacket place radiographs of assorted body parts and positions that have been properly marked and inaccurately marked. Instruct the students to evaluate the marker placement on the radiographs.

        Accuracy of radiation protection: In a jacket place radiographs of assorted body parts and positions that demonstrate accurate and inaccurate radiation protection placement. Instruct the students to evaluate the radiographs for proper radiation protection placement.

        Suggestion for learning activity number 12: Place radiographs demonstrating different types of artifacts in a jacket. Instruct the students to identify and categorize the artifacts.

    C. Other classroom activities.

        Filling out the requisition and repeat/reject analysis information: Provide your facility's uncompleted requisition and repeat/reject cards. Give the students information about an exam, such as film size and type, technique used, etc. and instruct them to use this information to complete the requisition and repeat/reject card.

        Judging the degree of part flexion: Place lateral knee, feet, and elbow radiographs that demonstrate varying degrees of part flexion. Instruct the students to state how much the part was flexed for each radiograph.

III. Clinical laboratory activities.

   A. Workbook learning activities numbers 4, 5, 6, 7, and 8.

   B. Other clinical activities.

   Marking of radiographs: Have students demonstrate and explain how to properly mark the film for hand, shoulder, chest, and abdominal radiographs. The AP/PA projections and oblique and lateral positions should be demonstrated. Describe how to determine where the marker is placed on a film that is in the bucky when the entire film size is not used. For example, where should the marker be placed if a 14 x 17 inch (35 x 43 cm) film is placed in the bucky, but only a 12 x 15 inch (30 x 37.5 cm) field size is used.

   Demonstrating how to measure patients when in a PA projection and lateral and oblique positions: Have a student position another student's hand in a PA projection and a lateral position. Center the central ray to the patient's metacarpal region and demonstrate how to measure the hand in each of these positions. Explain how snugly the calipers are placed against the patient and where to look on the calipers for the measurement. Demonstrate and discuss which technique chart is used, how to use the technique chart, and how to set the kVp and mAs on the generator for the measurements obtained.
   Position a student on the table in an AP projection and a lateral and a 45-degree oblique position. Center the central ray to the abdomen, then demonstrate how to measure the abdomen for each of these positions.

   Determining the source-image distance: Have the students demonstrate how to set the SID at 40 inches (102 cm) to the tabletop and to the bucky, when the central ray is perpendicular and when it is angled. Discuss what the resulting radiograph demonstrates if the SID is higher and lower than the routine.

   C. Workbook learning activities numbers 11 and 12.
   These activities are for radiography students that have been through a radiographic imaging class. I use these activities for senior students only.

IV. Final exam.
   The following questions pertain to the radiographic positioning and critiquing guidelines, and can be incorporated in your present positioning and critiquing exams. They are based on information found in the Radiographic Critique book. Since some facilities do not follow all the protocols found in the Radiographic Critique book, evaluate each question for information that does not match your facility's practices.

1. Which of the following are features of an optimal radiograph?
   1. Adequate penetration.
   2. Maximum recorded details.
   3. No motion.
   4. No removable artifacts.

   a. 1 and 2 only
   b. 1, 2, and 3 only
   c. 3 and 4 only
   *d. 1, 2, 3, and 4

2. Distal is a word used to refer to:
   a. a structure within the patient's torso that is situated closer to the feet.
   b. a structure of an extremity that is situated closest to the patient's torso.
   c. the patient's median plane.
   *d. a structure of an extremity that is situated farthest from the patient's torso.

3. Posterior is a word used to refer to:
   *a. the back surface of the patient.
   b. the front surface of the patient.
   c. the patient's median plane.
   d. the patient's sides.

4. Match the terms below with the definition that best describes it.
   1. __d__ foreshorten
   2. __e__ dorsiflex
   3. __h__ radiolucent
   4. __k__ cephalic
   5. __j__ flexion
   6. __i__ lateral rotation
   7. __m__ protract
   8. __c__ abduct
   9. __f__ convex
   10. __a__ extension
   11. __g__ caudal
   12. __n__ radiopaque
   13. __b__ retract
   14. __o__ pronate
   15. __l__ deviate

   a. Movement that straightens a structure.
   b. To move a structure posteriorly.
   c. To draw an extremity laterally.
   d. To make the long axis of a structure appear short on the radiographic image.
   e. To move toes and forefoot upward.
   f. Curved or rounded outward.
   g. The foot end of the patient.
   h. Allows x rays to pass through.
   i. Act of turning the anterior surface of an extremity away from the patient.
   j. A movement that bends a structure.
   k. The head end of the patient.
   l. To move away from the norm.
   m. To move a structure anteriorly.
   n. Prevents x-radiation from passing through.
   o. To rotate the upper extremity medially until the hand's palmar surface is posteriorly.

5. A PA projection of the chest is hung on the view box:
   1. with the marker correct.
   2. as if the patient is standing in an upright position.
   3. from the toes.
   4. with the marker reversed.

   a. 1 and 2 only
   b. 1 and 3 only
   *c. 2 and 4 only
   d. 3 and 4 only

6. A lateral foot radiograph is hung on the view box:
   1. with the marker correct.
   2. with the marker reversed.
   3. from the toes.
   4. as if from the patient's hip.

   a. 1 and 3 only
   *b. 1 and 4 only
   c. 2 and 3 only
   d. 2 and 4 only

7. An axiolateral shoulder radiograph is hung on the view box:
   1. with the anterior surface up.
   2. with the posterior surface up.
   3. with the marker correct.
   4. with the marker reversed.

   *a. 1 and 3 only
   b. 1 and 4 only
   c. 2 and 3 only
   d. 2 and 4 only

8. Which of the following is true about radiographic markers?
   1. They are radiopaque.
   2. They should be reversed before being placed on the radiographic cassette.
   3. Should be positioned as close to the median plane as possible.
   4. They will be magnified if positioned on the tabletop or patient.

   *a. 1 and 4 only
   b. 2 and 3 only
   c. 1, 2, and 4 only
   d. 3 and 4 only

9. Match the marker placement with the position/projection.

   1. __d__ lateral vertebrae   a. Laterally, on side being identified.
   2. __a__ PA cranium   b. Anteriorly, identifying side being radiographed.
   3. __e__ oblique vertebrae   c. Place marker anywhere within collimated field,
   4. __c__ lateral hand   identifying side being radiographed.
   5. __b__ crosstable hip   d. Anteriorly, indicating side placed adjacent to film.
                                   e. Laterally, identifying side situated closest to film.

10. When repositioning a patient for a repeat radiograph:
    1. start over trying to align the needed anatomy.
    2. move the patient the same distance demonstrated between where the mispositioned anatomical structures are and where they should be.
    3. begin by returning the patient to the original position before adjusting.
    4. move the patient half the distance demonstrated between where the mispositioned anatomical structures are and where they should be.

    a. 1 and 2 only
    b. 1 and 4 only
    c. 2 and 3 only
    *d. 3 and 4 only

11. A 15-degree caudally angled central ray was positioned above two superimposed anatomical structures. Which of these structures will be projected the most on the resulting radiograph?
    a. The structure positioned closest to the film.
    *b. The structure positioned farthest from the film.

12. Two anatomical structures that are superimposed on an accurately positioned radiograph are separated by 1 inch (2.5 cm). How much should the patient's positioning be adjusted before this radiograph is repeated?
    a. 0.25 inch (.6 cm)
    *b. 0.5 inch (1 cm)
    c. 1 inch (2.5 cm)
    d. 1.5 inches (4 cm)

13. Good collimation efforts:
    1. increase patient dosage.
    2. increase the visibility of recorded details.
    3. decrease the radiographic contrast.
    4. decrease the amount of scattered radiation that reaches the film.

    a. 1 and 3 only
    *b. 2 and 4 only
    c. 2 and 3 only
    d. 1, 2, 3, and 4

14. What gonadal organs should be shielded on a female patient?
    1. ovaries
    2. uterine tubes
    3. breast
    4. uterus

    a. 1 only
    b. 2 and 4 only
    *c. 1, 2, and 4 only
    d. 1, 2, 3, and 4

15. Which of the following contain cells that are radiosensitive.
    1. Breasts
    2. Eyes
    3. Thyroid
    4. Gonads

    a. 1 and 2 only
    b. 2 and 3 only
    c. 1, 3, and 4 only
    *d. 1, 2, 3, and 4

16. Which of the following are methods of controlling patient motion?
    1. Use a long exposure time.
    2. Use immobilization devices.
    3. Explain exam to patient.
    4. Make patient comfortable.

    a. 1 and 3 only
    b. 2 and 4 only
    *c. 2, 3, and 4 only
    d. 1, 2, 3, and 4

17. Motion on a radiograph that resulted when the patient yawned during the exposure is:
    *a. voluntary
    b. involuntary

18. A radiograph that demonstrates poor film-screen contact:
    1. demonstrates motion only where the screen and film were in poor contact.
    2. should be repeated.
    3. demonstrates blur throughout the radiograph.
    4. is acceptable.

    *a. 1 and 2 only
    b. 1 and 4 only
    c. 2 and 3 only
    d. 3 and 4 only

19. Two radiographs were taken of the same structure. Radiograph #1 was taken using a 72 inch (183 cm) SID and a 10 inch (25 cm) OID, and radiograph #2 was taken using a 72 inch (183 cm) SID and a 5 inch (13 cm) OID.
    Which radiograph demonstrates the greatest size distortion?
    *a. Radiograph #1
    b. Radiograph #2

    Which radiograph demonstrates the sharpest recorded details?
    a. Radiograph #1
    *b. Radiograph #2

20. A hypersthenic patient's thorax is:
    1. wide
    2. long
    3. narrow
    4. short

    a. 1 and 2 only
    *b. 1 and 4 only
    c. 2 and 3 only
    d. 3 and 4 only

21. Which technical factor listed below is primarily used to regulate density.
    a. kVp
    *b. mAs
    c. grids
    d. distances (SID, OID)

22. Which technical factor listed below is primarily used to regulate contrast.
    *a. kVp
    b. mAs
    c. grids
    d. distances (SID, OID)

23. Which of the following can control the amount of scattered radiation that reaches the film.
    1. Use a grid.
    2. Use a low mAs.
    3. Use tight collimation.
    4. Use a long SID.

    a. 1 only
    *b. 1 and 3 only
    c. 1, 3, and 4 only
    d. 2 and 4 only

24. Match the artifact with the artifact category it would fall under.

    1. __e__ anatomical artifact   a. grid lines
    2. __c__ external artifact   b. hesitation marks
    3. __d__ internal artifact   c. a safety pin
    4. __a__ equipment related artifact   d. total hip prosthesis
    5. __b__ improper film handling and processing   e. the patient's hand

Chapter 2

Radiographic Critique of the Chest and Abdomen

I. Lecture on radiographic projections/positions of the chest and abdomen.

II. Lecture on radiographic evaluation for each projection/position of the chest and abdomen.

III. Classroom activities.

   A. Workbook study questions.
      The workbook study questions are designed to encourage the student to form mental images of the radiographs and the patient positioning. It is this mental imaging that helps the students to understand the information rather than memorize it. The information is entirely

B. Workbook learning activities numbers 2, 3, 4, and 5.

Suggestion for learning activity numbers 2 and 3: I provide a skeleton that can be manipulated, and accurately and poorly positioned chest and abdominal radiographs. The students then work together, identifying the anatomical structures on the skeleton and on the radiographs.

This exercise helps the students to visualize how anatomical structures are magnified and distorted, determine which structures are superimposed, and how structures move in relationship to each other. Answers to questions in the workbook can also be discovered through this activity.

Suggestion for learning activity number 4: The class is broken into groups of 3 to 5 students. I have found it best to divide the groups so students of equal understanding are grouped together. This challenges the students with good understanding of radiographic positioning and evaluating to look more thoroughly at the radiograph and it prevents students with less understanding from depending on the better students to supply the answer. Each group is given a packet containing poorly positioned chest and abdominal radiographs. I have tried to make sure each problem listed in the Radiographic Critique book is demonstrated on at least one of the radiographs. Some of the radiographs have one problem and others have more than one problem.

If I am teaching junior students, the radiographs all have adequate density and contrast, since juniors have not studied this subject yet; and if I am teaching senior students, the radiographs have a combination of positioning, density, and contrast problems. The students are to identify the specific problem(s) found on the radiograph and explain how the patient was mispositioned for the radiograph or how the patient should be repositioned to obtain an optimal radiograph. I do not accept an answer if the students have listed how the exam is performed. I want a one to one correlation between each problem and each positioning procedure. (For a poorly positioned PA chest radiograph that demonstrates the scapulae in the lung field, the students response should be that the scapulae are in the lung field and the patient's arms and shoulders were not/should be rotated anteriorly.) I then correct the packets and make comments if anatomical terms were misused or the answer was incomplete. If a radiograph was inaccurately evaluated I do not give the answer, instead the packet is given back and the students are to reevaluate the radiograph. I of course am there to help answer any questions. I have found the students are less intimidated with this learning activity when it is not done for a grade.

Suggestion for learning activity number 5: I ask the students to each bring radiographs from the chest and abdomen repeat bin to class. During class they label all the anatomical structures found on the radiographs with a magic marker. These radiographs can later be returned to the film bin for silver collection. They are told to look for the radiographs a couple of weeks prior to the time the activity is done, so adequate radiographs can be found.

Doing activities that reinforce anatomy identification is essential. It is impossible to understand how structures have been accurately or inaccurately positioned, as described in the Radiographic Critique book, if one does not know where the structures are located.

IV. Clinical laboratory activities.

A. Workbook learning activity numbers 1.

Suggestions for learning activity number 1: I assign each student a chest or abdomen projection/position to demonstrate to the class. They are aware of the position they must be prepared to demonstrate at least a couple of weeks before the laboratory class, and the positioning procedure and radiographic evaluation has been thoroughly discussed in the classroom setting prior to the laboratory class. During the laboratory session they are to place another student into the assigned position. While doing so they are to explain why each step is performed (i.e., the patient's shoulders and elbows are rotated anteriorly for a PA chest to remove the scapulae out of the lung field), and be prepared to answer any questions about patient and central ray positioning and radiographic evaluation that other students and I might ask. The questions asked, range from why a certain procedure is followed for this position to what would the radiograph show if the patient was positioned a different way. I grade the student's performance on this laboratory activity much the same as our program grades the positioning clinical competencies. Throughout the semester each student demonstrates many different positions. The grade obtained from this section is 10 percent of the final positioning grade.

V. Workbook self test questions.

Suggestion: I ask the students not to look at or do the self test prior to completing the study questions and the learning activities. After the study questions and learning activities are completed the self test is taken in the classroom setting as if it were a quiz.
Used in this manner the self test is a tool that pinpoints areas of weakness before the final exam.

VI. Final Exam.
The following are questions that pertain to both radiographic positioning and critique, and can be incorporated in your present positioning and critiquing exams. They are based on information found in the Radiographic Critique book. Since some facilities do not follow all of the protocols found in the Radiographic Critique book, evaluate each question for information that does not match your facility's routines.

1. For an accurately positioned PA chest radiograph:
   1. the SID is set at 72 inches (183 cm).
   2. the shoulders are positioned at equal distances to the film.
   3. the upper midcoronal plane is tilted slightly toward the film.
   4. the elbows and shoulders are rotated posteriorly.

   *a. 1 and 2 only
   b. 2 and 3 only
   c. 1, 2, and 4 only
   d. 1, 2, 3, and 4

2. An accurately positioned PA chest radiograph demonstrates:
   1. 10 to 11 posterior ribs above the diaphragm.
   2. the air-filled trachea aligned with the vertebral column.
   3. the manubrium superimposed by the fourth thoracic vertebra.
   4. the scapulae outside the lung field.

   a. 1, 2, and 4 only
   b. 2 and 4 only
   c. 1 and 3 only
   *d. 1, 2, 3, and 4

3. The film is positioned _____ for a PA chest radiograph taken on a hypersthenic patient.
   *a. crosswise
   b. lengthwise

4. A PA chest radiograph taken on expiration demonstrates:
   1. A narrower and longer heart shadow.
   2. An underexposed radiograph if exposure is not increased.
   3. A broader and shorter heart shadow
   4. Less than 10 posterior ribs above the diaphragm.

   a. 1 and 2 only
   b. 1, 2, and 4 only
   c. 3 and 4 only
   *d. 2, 3, and 4 only

5. Upon inhalation the lungs expand:
   1. Vertically
   2. Transversely
   3. Anteroposteriorly

   a. 1 only
   b. 2 and 3 only
   c. 1 and 2 only
   *d. 1, 2, and 3

6. A PA chest radiograph is taken with the patient rotated into an RAO position demonstrates which of the following?
   1. 1 inch (2.5 cm) of the apical lung field visualized above the clavicle.
   2. The vertebral column superimposes the right SC joint.
   3. Elevated lateral clavicular ends.
   4. An uneven density between the lateral borders of the chest.

   a. 2 only
   b. 1, 2, and 4 only
   c. 3 only
   d. 2 and 4 only

7. A poorly positioned PA chest radiograph demonstrates vertical clavicles and the manubrium at the same level as the 5th thoracic vertebra. How was the patient mispositioned to obtain such a radiograph?
   a. The shoulders and elbows were not internally rotated.
   b. The shoulders were elevated.
   *c. The patient's upper midcoronal plane was tilted toward the film.
   d. The central ray was angled caudally.

8. A poorly positioned PA chest radiograph demonstrates the scapulae in the lung field and elevated lateral clavicular ends. How should the patient be repositioned to obtain an optimal radiograph?
   1. Tilt the upper midcoronal plane away from the film.
   2. Depress the shoulders.
   3. Coax the patient into a deeper inspiration.
   4. Anteriorly rotate the shoulders and elbows.

   a. 1 and 4 only
   b. 2 only
   c. 2, 3, and 4 only
   *d. 2 and 4 only

9. For an accurately positioned lateral chest radiograph:
   1. the SID is set at 40 inches (102 cm).
   2. the humeri are positioned at a 90-degree angle with the body.
   3. the shoulders, the posterior ribs, and the posterior pelvic wings are aligned perpendicular to the film.
   4. the midsagittal plane is aligned perpendicular to the film.

   a. 1 and 3 only
   b. 2 and 4 only
   *c. 2 and 3 only
   d. 3 and 4 only

10. An accurately positioned lateral chest radiograph demonstrates:
    1. no humeral soft tissue demonstrated in the lung field.
    2. no more than 0.5 inch (1 cm) of space between the posterior ribs.
    3. the right hemidiaphragm inferior to the left hemidiaphragm.
    4. the hemidiaphragms inferior to the 11th thoracic vertebra.

    *a. 1, 2, and 4 only
    b. 1 and 2 only
    c. 3 and 4 only
    d. 2 and 3 only

11. The last rib is attached to the _____ vertebra.
    a. eleventh
    b. tenth
    *c. twelfth
    d. ninth

12. Which of the following pertains to an accurately positioned lateral chest radiograph taken with the right side positioned adjacent to the film?
    1. The heart shadow demonstrates increased magnification.
    2. The left lung demonstrates the sharpest recorded details.
    3. The left hemidiaphragm is demonstrated inferior to the right hemidiaphragm.
    4. There are 1.5 inches (4 cm) of space between the posterior ribs.

    a. 1 only
    *b. 1 and 3 only
    c. 1, 2, and 3 only
    d. 2 and 4 only

13. A lateral chest radiograph taken with the patient's left side rotated anteriorly demonstrates:
    1. the anterior and the posterior ribs without superimposition.
    2. the heart shadow anterior to the sternum.
    3. the right hemidiaphragm inferior to the left hemidiaphragm.
    4. the humeral soft tissue superimposing the anterior lung apices.

    *a. 1 and 2 only
    b. 3 and 4 only
    c. 1, 2, and 4 only
    d. 1, 3, and 4 only

14. A poorly positioned lateral chest radiograph demonstrates the humeri soft tissue superimposing the anterior lung apices. How was the patient positioned to obtain such a radiograph?
    a. The chest was rotated.
    b. The lower midsagittal plane was tilted toward the film.
    c. The humeri were positioned at a 90-degree angle with the body.
    d. The central ray was angled caudally.

15. A lateral chest radiograph demonstrates the gastric air bubble directly beneath the superior hemidiaphragm. Identify the superior lung.
    *a. left
    b. right

16. A rotated lateral chest radiograph demonstrates the heart shadow posterior to the sternum. Identify the anteriorly positioned lung.
    a. left
    *b. right

17. For an AP chest radiograph taken with a portable machine:
    1. the film is positioned parallel with the patient's bed.
    2. the radiograph is taken without using a grid.
    3. the manubrium superimposes the fourth thoracic vertebra.
    4. 10 to 11 posterior ribs are demonstrated above the diaphragm.

    a. 1 and 2 only
    *b. 1, 2, and 3 only
    c. 3 and 4 only
    d. 1, 2, 3, and 4

18. Air-fluid levels on an AP chest radiograph:
    1. demonstrate a increase in density where the fluid is present.
    2. are formed when air and fluid within the chest cavity separate.
    3. are precisely demonstrated when the patient is in a partially upright position.
    4. are precisely demonstrated when the central ray is horizontal.

    a. 1 and 3 only
    *b. 1, 2, and 4 only
    c. 2 and 4 only
    d. 2, 3, and 4 only

19. Heart penetration on an AP chest radiograph:
    1. is obtained by increasing the kVp.
    2. results in a lower contrast radiograph.
    3. is required when apparatuses located at the mediastinal region are of interest.
    4. results in a decrease in scattered radiation reaching the film.

    a. 1 and 3 only
    *b. 1, 2, and 3 only
    c. 3 only
    d. 1 and 4 only

20. A portable AP chest radiograph taken with the central ray angled caudally demonstrates:
    1. vertically contoured ribs.
    2. the manubrium projected superior to the fourth thoracic vertebra.
    3. less than 1 inch (2.5 cm) of the apices above the clavicles.
    4. vertical clavicles.

    a. 1 and 3 only
    b. 2 and 3 only
    c. 2 and 4 only
    *d. 1 and 4 only

21. An AP chest radiograph taken with the patient rotated into an RPO position demonstrates:
    1. the left SC joint superimposing the vertebral column.
    2. the right SC joint superimposing the vertebral column.
    3. an uneven density between the lateral borders of the chest.
    4. elevated lateral clavicular ends.

    *a. 1 and 3 only
    b. 1 and 4 only
    c. 2 and 3 only
    d. 2 and 4 only

22. For a lateral decubitus chest radiograph:
    1. the shoulders and the posterior ribs are positioned perpendicular to the cart.
    2. the humeri are positioned at a 90-degree angle with the film.
    3. the midcoronal plane is aligned perpendicular with the film.
    4. the patient is elevated on a radiolucent sponge or cardiac board.

    *a. 1 and 4 only
    b. 2 and 3 only
    c. 1, 2, and 4 only
    d. 1, 2, 3, and 4

23. An accurately positioned lateral decubitus chest radiograph demonstrates:
    1. an arrow marker pointing upward to indicate the side adjacent to the cart.
    2. the air-filled trachea aligned with the vertebral column.
    3. the manubrium at the fifth thoracic vertebra.
    4. nine to ten posterior ribs above the diaphragm.

    a. 1, 2, and 4 only
    *b. 2 and 4 only
    c. 2 only
    d. 3 and 4 only

24. Which side of the patient is positioned against the tabletop or cart for a lateral decubitus chest radiograph to rule-out a right pneumothorax.
    a. right
    *b. left

25. Which side of the patient is positioned against the tabletop or cart for a lateral decubitus chest radiograph to rule-out a left side pleural effusion.
    a. right
    *b. left

26. A lateral decubitus chest radiograph taken in a PA projection demonstrates:
    1. the C6-7 vertebral bodies without distortion.
    2. the manubrium superimposing the fourth vertebral body.
    3. a closed C6-7 intervertebral disk space.
    4. well visualized C6-7 spinous processes and laminae.

    a. 1 and 2 only
    b. 2 and 3 only
    *c. 2, 3, and 4 only
    d. 3 and 4 only

27. A lateral decubitus chest radiograph taken with the patient in an RPO position demonstrates:
    1. the right SC joint without vertebral column superimposition.
    2. 9 to 10 posterior ribs above the diaphragm.
    3. the manubrium superimposing the first thoracic vertebra.
    4. the vertebral column superimposing the left SC joint.

    a. 1 and 2 only
    *b. 1, 2, and 4 only
    c. 2 and 3 only
    d. 1 and 4 only

28. Which positioning problem(s) listed below result in a lateral decubitus chest radiograph with the manubrium and the fifth thoracic vertebra located at the same level?
    1. A patient rotated into an RPO position.
    2. An AP projection taken with the upper midcoronal plane tilted away from the film.
    3. A PA projection taken with the upper midcoronal plane tilted toward the film.
    4. A patient whose humeri are elevated above the head.

    a. 1, 2, and 3 only
    b. 2 only
    *c. 2 and 3 only
    d. 3 and 4 only

29. For an AP lordotic chest radiograph:
    1. the shoulders are positioned at equal distances to the film.
    2. the patient's back is arched until the midcoronal plane and film form a 45-degree angle.
    3. a 15-degree caudal central ray angulation is used.
    4. the elbows and shoulders are rotated anteriorly.

    a. 1 and 2 only
    b. 2 and 3 only
    *c. 1, 2, and 4 only
    d. 1, 2, 3, and 4

30. An accurately positioned AP lordotic chest radiograph demonstrates:
    1. the medial ends of the clavicles projected superior to the lung apices.
    2. the lateral borders of the scapulae within the lung field.
    3. equal distances from the vertebral column to the SC joints.
    4. nearly horizontal posterior and anterior portions of the first through fourth ribs.

    a. 1 and 3 only
    b. 1 and 4 only
    c. 2 and 3 only
    *d. 1, 3, and 4 only

31. A poorly positioned lordotic chest radiograph demonstrates the clavicles within the lung apices. How should the positioning setup be adjusted to obtain an optimal radiograph?
    1. Increase the degree of cephalic central ray angulation.
    2. Anteriorly rotate the elbows and shoulders.
    3. Arch the patient's back more, increasing the midcoronal plane to film angle.
    4. Position the patient's feet closer to the film.

    *a. 1 and 3 only
    b. 2 only
    c. 1, 2, and 3 only
    d. 1, 2, 3, and 4

32. For an anterior oblique chest radiograph:
    1. the patient is rotated until the midsagittal plane is aligned 45 degrees with the film.
    2. there is twice as much lung field demonstrated on one side of the vertebral column than the opposite side.
    3. 10 to 11 posterior ribs are demonstrated above the hemidiaphragm.
    4. the apices, costophrenic angles, and lateral chest walls are included on the radiograph.

    a. 1 and 4 only
    b. 2 and 3 only
    *c. 2, 3, and 4 only
    d. 1, 2, 3, and 4

33. An RAO chest position corresponds with what posterior oblique position?
    a. RPO
    *b. LPO

34. An LAO 60-degree oblique chest radiograph:
    1. demonstrates the heart shadow to the right of the vertebral column.
    2. is taken to evaluate the size and configuration of the heart shadow.
    3. best demonstrates the left lung.
    4. demonstrates more than twice as much lung field on one side of the vertebral column than the opposite side.

    a. 1 and 2 only
    b. 1 and 3 only
    *c. 1, 2, and 4 only
    d. 2 and 4 only

35. A poorly positioned RAO chest radiograph demonstrates nearly equal lung field on each side of the vertebral column and nine posterior ribs above the diaphragm. How could the positioning setup be adjusted to obtain an optimal radiograph?
    1. Take the exposure after the second full inspiration.
    2. Increase the degree of patient obliquity.
    3. Decrease the degree of patient obliquity.
    4. Move the film and central ray inferiorly.

    *a. 1 and 2 only
    b. 1 and 3 only
    c. 2 and 4 only
    d. 1 and 4 only

36. For an upright AP abdominal radiograph:
    1. the ASISs are positioned at equal distances from the film.
    2. the patient remains in an upright position at least 10 minutes prior to taking the radiograph.
    3. the symphysis pubis and obturator foramina are included.
    4. the patient is instructed to take a deep inspiration prior to taking the radiograph.

    *a. 1 and 2 only
    b. 1, 2, and 3 only
    c. 2 and 3 only
    d. 3 and 4 only

37. An accurately positioned supine abdominal radiograph demonstrates:
    1. the outline of the psoas muscles and kidneys.
    2. the symphysis pubis and obturator foramina.
    3. the spinous processes aligned with the midline of the vertebral bodies.
    4. the long axis of the vertebral column aligned with the long axis of the collimated field.

    a. 1 and 2 only
    b. 3 and 4 only
    c. 1, 2, and 3 only
    *d. 1, 2, 3, and 4

38. How much should the technique be adjusted from the routine for an AP abdominal radiograph on a patient that has large amounts of bowel gas?
    1. Increase the mAs 30 to 50 percent.
    2. Decrease the mAs 30 to 50 percent.
    3. Increase the kVp 5 to 8 percent.
    4. Decrease the kVp 5 to 8 percent.

    a. 1 and 3 only
    b. 1 and 4 only
    c. 2 and 3 only
    *d. 2 and 4 only

39. Voluntary motion:
    1. can result from patient breathing.
    2. can be controlled by using a short exposure time.
    3. can result from peristaltic activity.
    4. can be identified as sharp bony cortices and blurry gastric and intestinal gases.

    *a. 1 and 2 only
    b. 1, 2, and 4 only
    c. 2 and 3 only
    d. 3 and 4 only

40. Which of the following is true about an AP abdominal radiograph taken on a patient who was in an LPO position?
    1. The sacrum and coccyx are aligned.
    2. The distance from the pedicles to the spinous processes on the right side is narrower than the same distance on the left side.
    3. The sacrum is closer to the patient's right side.
    4. The symphysis pubis is rotated toward the patient's right side.

    a. 1, 2, and 3 only
    *b. 2 and 3 only
    c. 2 and 4 only
    d. 1 and 4 only

41. For a lateral decubitus abdominal radiograph:
    1. the right hemidiaphragm and iliac wing must be included to demonstrate intraperitoneal air.
    2. position the shoulders and the ASISs at equal distances to the tabletop.
    3. take the exposure on expiration.
    4. position the patient's right side adjacent to the tabletop or cart.

    a. 1 and 2 only
    *b. 1 and 3 only
    c. 2 and 4 only
    d. 1, 2, and 4 only

42. How should the technique be adjusted from the routine for a decubitus chest radiograph on a patient with ascites or a bowel obstruction?
    1. Increase mAs by 30 to 50 percent.
    2. Decrease mAs by 30 to 50 percent.
    3. Increase kVp 5 to 8 percent.
    4. Decrease kVp 5 to 8 percent.

    *a. 1 and 3 only
    b. 1 and 4 only
    c. 2 and 3 only
    d. 2 and 4 only

Chapter 3
Radiographic Critique of the Upper Extremity

I. Lecture on radiographic projections/positions of the upper extremity.

II. Lecture on radiographic evaluation for each projection/position of the upper extremity.

III. Classroom activities.

   A. Workbook study questions.

   B. Workbook learning activities numbers 2, 3 and 4.

   Suggestion for learning activity number 2: I provide several connected, movable upper extremity bones, and accurately and poorly positioned upper extremity radiographs. The students then work together, identifying the anatomical structures on the bones and on the radiographs.

   Suggestion for learning activity number 3: See chapter 2, suggestion for learning activity number 4, page 7. Fill the packets with poorly positioned upper extremity radiographs.

   Suggestion for learning activity number 4: I ask the students to each bring radiographs from the upper extremity repeat bin to class. During class they label all the anatomical structures found on the radiographs with a magic marker.

IV. Clinical laboratory activities.
   A. Workbook learning activity number 1.

   Suggestion for learning activity number 1: I assign each student an upper extremity projection/position to demonstrate to the class. See chapter 2, suggestions for learning activity number 1, page 7 for further description.

V. Workbook self test questions.

   Suggestion: I ask the students not to look at or do the self test prior to completing the study questions and learning activities. After the study questions and learning activities are completed the self test is taken in the classroom setting as if it were a quiz.
   Used in this manner the self test is a tool that pinpoints areas of weakness before the final exam.

VI. Final exam.
   The following questions pertain to both radiographic positioning and critique, and can be incorporated in your present positioning and critiquing exams. They are based on information found in the Radiographic Critique book. Since some facilities do not follow all the protocols found in the Radiographic Critique book, evaluate each question for information that does not match your facility's routines.

1. Sharply recorded details are demonstrated on extremity radiographs when:
   1. motion is controlled.
   2. a large focal spot is used.
   3. a detail screen is used.
   4. a large OID is used.

   *a. 1 and 3 only
   b. 2 and 4 only
   c. 1, 3, and 4 only
   d. 1, 2, 3, and 4

2. Match one of the following with the position/projection listed below.
   a. Phalanges demonstrate equal concavity.
   b. Phalanges demonstrate more concavity on one side than on the other.
   c. Phalanges demonstrate concavity on one side and convexity on the other.

   __c__ 1. Lateral finger position
   __a__ 2. PA finger projection
   __b__ 3. Oblique finger position

3. The interphalangeal joint spaces on a finger radiograph are open when:
   1. the central ray is aligned parallel with them.
   2. the central ray is aligned perpendicular to them.
   3. the joints are aligned parallel with the film.
   4. the joints are aligned perpendicular to the film.

   a. 1 and 3
   *b. 1 and 4
   c. 2 and 3
   d. 2 and 4

4. List how the thumb is placed for each of the following positions/projections?
   a. In an oblique position.
   b. In a lateral position.
   c. In an AP projection.

   __a__ 1. Hand is pronated
   __b__ 2. Hand is flexed
   __c__ 3. Hand is internally rotated

5. Which side of the arm is positioned against the film for the following lateral finger radiographs?
   a. ulnar
   b. radial

   __b__ 1. Second finger
   __a__ 2. Fourth finger

6. Which of the following technical factors should be chosen when 20 mAs is desired and the patient being radiographed has difficulty holding still?

   a. 200 mA @ 0.1 sec.
   *b. 400 mA @ 0.05 sec.
   c. 100 mA @ 0.4 sec.
   d. 100 mA @ 0.2 sec

7. A lateral finger radiograph taken with the finger in an oblique position demonstrates:
   1. equal soft tissue width on each side of the phalanges.
   2. more midshaft concavity on one side of the phalanges than the opposite side.
   3. twice as much soft tissue on one side of the phalanges than the opposite side.
   4. convexity on one side of the phalanges and concavity on the opposite side.

   a. 1 and 2 only
   b. 1 and 4 only
   *c. 2 and 3 only
   d. 3 and 4 only

8. A PA hand radiograph taken with the hand flexed demonstrates:
   1. foreshortened phalanges.
   2. the thumb in a lateral position.
   3. closed interphalangeal joint spaces.
   4. foreshortened metacarpals.

   a. 1 and 3 only
   b. 2 and 4 only
   c. 1, 3, and 4 only
   *d. 1, 2, 3, and 4

9. A lateral hand radiograph taken with the hand in slight external rotation demonstrates:
   1. the shortest metacarpal anterior to the other metacarpals.
   2. the radius posterior to the ulna.
   3. the second metacarpal posterior to the other metacarpals.
   4. the pisiform posterior to the distal scaphoid.

   a. 1 and 2 only
   *b. 1, 2, and 3 only
   c. 3 and 4 only
   d. 1, 2, and 4 only

10. State where the soft tissue structures that can be used to indicate joint effusion are located on the following projection/positions.
    a. Anterior
    b. Medially
    c. Laterally
    d. Posteriorly

    __a__ 1. Lateral wrist
    __c__ 2. PA wrist
    __c__ 3. Scaphoid wrist

11. An accurately positioned PA wrist radiograph demonstrates:
    1. an open radioulnar joint.
    2. the radial styloid in profile.
    3. the ulnar styloid projected distal to the midline of the ulnar head.
    4. superimposition of the third through fifth metacarpal bases.

    *a. 1 and 2 only
    b. 1, 2, and 3 only
    c. 3 and 4 only
    d. 1, 3, and 4 only

12. How is a patient positioned for a PA wrist radiograph to superimpose the anterior and posterior margins of the distal radius and obtain open radioscaphoid and radiolunate joint spaces?
    a. Align the third metacarpal with the midforearm.
    b. Ulnar flex the wrist.
    *c. Elevate the proximal forearm.
    d. Flex the hand until the metacarpals are at a 10 to 15 degree angle with the film.

13. A PA wrist radiograph taken in slight external rotation demonstrates:
    1. superimposition of the laterally located carpal bones.
    2. a closed radioulnar joint.
    3. open lateral carpal joint spaces.
    4. the radial styloid in profile.

    a. 1 and 4 only
    b. 2 and 3 only
    *c. 2, 3, and 4 only
    d. 1, 2, and 4 only

14. A PA wrist radiograph taken with the hand flexed and the metacarpals at a 45-degree angle with the film demonstrates:
    1. the ulnar styloid projected distal to the ulnar head midline.
    2. foreshortened metacarpals.
    3. a decrease in scaphoid foreshortening.
    4. closed second through fifth carpometacarpal joint spaces.

    a. 1 and 3 only
    b. 1, 2, and 4 only
    *c. 2, 3, and 4 only
    d. 1, 2, 3, and 4 only

15. A PA wrist radiograph taken with the wrist in a neutral position demonstrates:
    1. the scaphoid in partial foreshortening.
    2. the center of the lunate positioned distal to the radioulnar articulation.
    3. closed CM joints.
    4. alignment of the long axis of the third metacarpal and radius.

    a. 1 and 3 only
    b. 2 and 4 only
    *c. 1, 2, and 4 only
    d. 1, 2, 3, and 4

16. A PA wrist radiograph taken in radial flexion demonstrates:
    1. the lunate positioned distal to the ulna.
    2. a foreshortened scaphoid.
    3. closed carpometacarpal joints.
    4. an elongated scaphoid.

    *a. 1 and 2 only
    b. 3 and 4 only
    c. 1, 3, and 4 only
    d. 1, 2, and 3 only

17. An accurately positioned medially obliqued wrist radiograph demonstrates:
    1. the trapezoid and trapezium without superimposition.
    2. an open radioulnar joint space.
    3. the ulnar styloid in profile.
    4. superimposition of the medially located carpals.

    a. 2 and 3 only
    b. 2 and 4 only
    *c. 1, 3, and 4 only
    d. 1, 2, 3, and 4 only

18. A lateral wrist radiograph taken with the elbow flexed 90 degrees and humerus placed parallel with the film demonstrates:
    1. the ulnar styloid distal to the ulnar head's midline.
    2. superimposition of the radius and ulna.
    3. superimposition of the distal scaphoid and pisiform.
    4. the ulnar styloid in profile.

    a. 1 and 2 only
    b. 3 and 4 only
    c. 1, 2, and 3 only
    *d. 2, 3, and 4 only

19. The trapezium is demonstrated on a lateral wrist radiograph when the patient:
    a. positions the wrist in slight internal rotation.
    b. hyperextends the wrist.
    *c. depresses the distal first metacarpal.
    d. ulnar flexes the wrist.

20. A lateral wrist radiograph taken with the wrist in slight internal rotation demonstrates
    1. The distal scaphoid anterior to the pisiform.
    2. The radius posterior to the ulna.
    3. The distal scaphoid distal to the pisiform.
    4. The radius anterior to the ulna.

    a. 1 and 2 only
    b. 3 and 4 only
    *c. 1 and 4 only
    d. 2 and 3 only

21. For each of the following situations match the central ray angulation that is used for PA, ulnar flexed wrist radiograph.
    a. 15 degrees
    b. 20 degrees
    c. 5 to 10 degrees
    d. 20 to 25 degrees

    __b__ 1. Patient is unable to ulnar flex wrist.
    __a__ 2. Patient ulnar flexed until first metacarpal and radius are aligned.
    __c__ 3. To evaluate a proximal scaphoid fracture.
    __d__ 4. To evaluate a distal scaphoid fracture.

22. When the patient ulnar flexes for a PA, ulnar flexed wrist radiograph:
    1. the first metacarpal and radius are aligned.
    2. the distal scaphoid shifts anteriorly.
    3. the lunate is demonstrated distal to the radius.
    4. the distal scaphoid shifts posteriorly.

    a. 1 and 2 only
    b. 3 and 4 only
    *c. 1, 3, and 4 only
    d. 1, 2, and 3 only

23. To take advantage of the anode-heel affect on a forearm radiograph:
    a. a detail screen is used.
    b. the elbow is positioned at the anode end of the x-ray tube.
    *c. the wrist is positioned at the anode end of the x-ray tube.
    d. a 55 to 65 kVp is used.

24. A film that is large enough to extend at least 1 inch (2.5 cm) beyond the elbow and wrist joints for a forearm radiograph:
    *a. is needed so the diverged x rays used to record the elbow and wrist are included on radiograph.
    b. is needed so the beam can be tightly collimated.
    c. is not a required technique.
    d. is needed so none of the film is wasted.

25. An accurately positioned AP forearm radiograph demonstrates:
    1. the radial styloid in profile laterally.
    2. the radial head and radial tuberosity superimposing the ulna by 0.25 inch (.6 cm).
    3. the ulnar styloid in profile medially.
    4. the humeral epicondyles in profile.

    a. 1 and 3 only
    b. 2, 3, and 4 only
    *c. 1, 2, and 4 only
    d. 1, 2, 3, and 4 only

26. An AP forearm radiograph taken with the wrist and elbow in lateral rotation demonstrates:
    1. superimposed first and second metacarpal bases.
    2. the proximal radius superimposes the ulna by more than 0.25 inches (.6 cm).
    3. superimposed fourth and fifth metacarpal bases.
    4. the proximal radius and ulna without superimposition.

    a. 1 and 2 only
    b. 1 and 4 only
    c. 2 and 3 only
    *d. 3 and 4 only

27. For a lateral forearm radiograph:
    1. the long axis of the forearm is aligned with the long axis of the collimated field.
    2. the patient's wrist is positioned at the anode end of the tube.
    3. the elbow is placed in a lateral position.
    4. a detail screen is used.

    a. 1 and 4 only
    b. 1, 3, and 4 only
    c. 2 and 3 only
    *d. 1, 2, 3, and 4

28. Which of the following projections is taken to prevent crossing of the forearm bones?
    *a. AP projection
    b. PA projection

29. An accurately positioned lateral forearm radiograph demonstrates:
    1. the distal scaphoid slightly distal to the pisiform.
    2. the ulnar styloid in profile.
    3. an open elbow joint space.
    4. the radial tuberosity in profile.

    *a. 1, 2, and 3 only
    b. 1 and 3 only
    c. 2 and 4 only
    d. 1, 2, 3, and 4 only

30. A lateral forearm radiograph taken on a patient with the proximal humerus elevated and the wrist internally rotated demonstrates:
    1. the radial head posterior to the coronoid process.
    2. the pisiform anterior to the distal scaphoid.
    3. the capitulum distal to the medial trochlea.
    4. the pisiform distal to the distal scaphoid.

    a. 1 and 2 only
    b. 1 and 4 only
    *c. 2 and 3 only
    d. 3 and 4 only

31. An accurately positioned AP elbow radiograph demonstrates:
    1. the medial and lateral humeral epicondyles in profile.
    2. the radial tuberosity in profile medially.
    3. an open capitulum-radial joint.
    4. the ulna free of radial head and radial tuberosity superimposition.

    a. 1 and 3 only
    b. 2 and 4 only
    *c. 1, 2, and 3 only
    d. 1, 2, 3, and 4 only

32. An AP elbow radiograph taken with the elbow medially rotated may demonstrate:
    1. the radius crossing over the ulna.
    2. an open capitulum-radial joint space.
    3. more than 0.25 inch (.6 cm) of proximal radial and ulnar superimposition.
    4. less than 0.25 inch (.6 cm) of proximal radial and ulnar superimposition.

    a. 1 and 3 only
    *b. 2 and 3 only
    c. 2 and 4 only
    d. 1, 2, and 3 only

33. For an accurately positioned laterally obliqued elbow radiograph:
    1. the capitulum in profile.
    2. an open capitulum-radial joint space.
    3. the coronoid process in profile.
    4. the ulna without radial head superimposition.

    *a. 1, 2, and 4 only
    b. 2, 3, and 4 only
    c. 2 and 3 only
    d. 1 and 4 only

34. An accurately positioned medially obliqued elbow radiograph demonstrates which of the following structures in profile?
    1. Capitulum
    2. Radial head
    3. Medial trochlea
    4. Coronoid process

    a. 1 and 2
    b. 1 and 4
    c. 2 and 3
    *d. 3 and 4

35. An accurately positioned lateral elbow radiograph demonstrates:
    1. an open elbow joint space.
    2. the radial head distal to the coronoid process.
    3. the radial tuberosity out of profile.
    4. the anterior fat pad.

    a. 2 and 3 only
    b. 1 and 4 only
    *c. 1, 3, and 4 only
    d. 1, 2, 3, and 4

36. A lateral elbow radiograph taken with the wrist and hand pronated demonstrates:
    1. the radial head anterior to the coronoid.
    2. the radial tuberosity in profile anteriorly.
    3. an open capitulum-radial joint.
    4. the radial tuberosity in profile posteriorly.

    a. 2 only
    *b. 4 only
    c. 1, 3, and 4 only
    d. 1 and 3 only

37. A lateral elbow radiograph demonstrates the radial head situated anterior and proximal to the coronoid process. How was the patient positioned to obtain such a radiograph?
    1. Distal forearm was too high.
    2. Distal forearm was too low.
    3. Proximal forearm was too high.
    4. Proximal forearm was too low.

    a. 1 and 3 only
    *b. 1 and 4 only
    c. 2 and 3 only
    d. 2 and 4 only

38. A lateral elbow radiograph that was taken with the distal forearm position too low and the proximal humerus position too high demonstrates:
    1. the radial head distal and posterior to the coronoid process.
    2. the radial head proximal and anterior to the coronoid process.
    3. the capitulum posterior and proximal to the medial trochlea.
    4. the capitulum anterior and distal to the medial trochlea.

    a. 1 and 3 only
    *b. 1 and 4 only
    c. 2 and 3 only
    d. 2 and 4 only

Chapter 4

Radiographic Critique of the Shoulder

I. Lecture on radiographic projections/positions of the shoulder.

II. Lecture on radiographic evaluation for each projection/position of the shoulder.

III. Classroom activities.

    A. Workbook study questions.

    B. Workbook learning activities numbers 2, 3, and 4.

    Suggestion for learning activity number 2: I provide connected and unconnected humeral and scapular bones, and accurately and poorly positioned shoulder radiographs. The students then work together, identifying the anatomical structures on the bones and on the radiographs.

    Suggestion for learning activity number 3: See chapter 2, suggestion for learning activity number 4, page 7. Fill the packets with poorly positioned shoulder radiographs.

    Suggestion for learning activity number 4: I ask the students to each bring radiographs from the shoulder repeat bin to class. During class they label all the anatomical structures found on the radiographs.

IV. Clinical laboratory activities.

  A. Workbook learning activities number 1.

   Suggestion for learning activity number 1: I assign each student a shoulder projection/position to demonstrate to the class. See chapter 2, suggestions for learning activity number 1, page 7 for further description.

V. Workbook self test questions.

  Suggestion: I ask the students not to look at or do the self test prior to completing the study question and learning activities. After the study questions and learning activities are completed the self test is taken in the classroom setting as if it were a quiz.
  Used in this manner the self test is a tool that pinpoints areas of weakness before the final exam.

VI. Final exam.
  The following questions pertain to both radiographic positioning and critique, and can be incorporated in your present positioning and critiquing exams. They are based on information found in the Radiographic Critique book. Since all facilities do not follow all the protocols found in the Radiographic Critique book, evaluate each question for information that does not match your facility's routines.

1. An accurately positioned AP shoulder radiograph demonstrates:
   1. the glenoid fossa in profile.
   2. the glenohumeral joint centered within the collimated field.
   3. the superolateral scapular border without thorax superimposition.
   4. the medial clavicular end positioned next to the vertebral column.

   a. 1, 2, 3 only
   *b. 2, 3, and 4 only
   c. 2 and 4 only
   d. 3 and 4 only

2. An AP right shoulder radiograph taken with the patient's body rotated away from the right shoulder demonstrates:
   1. the scapula with decreased thorax superimposition.
   2. the medial end of the right clavicle superimposing the vertebral column.
   3. a transversely foreshortened scapular body.
   4. the glenoid fossa in profile.

   a. 1 and 4 only
   *b. 1, 2, and 3 only
   c. 3 and 4 only
   d. 1, 2, 3, and 4

3. An AP shoulder radiograph demonstrates longitudinal foreshortening of the scapular body when:
   1. the patient's upper midcoronal plane is tilted away from the film.
   2. the patient is rotated onto the affected shoulder.
   3. the patient is kyphotic.
   4. the affected shoulder is protracted.

   *a. 1 and 3 only
   b. 2 and 4 only
   c. 1, 3, and 4 only
   d. 1, 2, 3, and 4

4. An AP shoulder radiograph taken with the humeral epicondyles positioned parallel with the film demonstrates:
   1. the greater tubercle in profile laterally.
   2. the lesser tubercle in profile medially.
   3. the lesser tubercle demonstrated halfway between the greater tubercle and medial aspect of the humeral head.
   4. the greater tubercle superimposing the humeral head.

   *a. 1 and 3 only
   b. 1 and 2 only
   c. 3 and 4 only
   d. 2 and 4 only

5. For an AP shoulder radiograph:
   1. the shoulders are positioned at equal distances from the film.
   2. the central ray is centered to the coracoid.
   3. an imaginary line connecting the humeral epicondyles is positioned at a 45 degree angle with the film.
   4. the central ray is angled cephalically when radiographing a kyphotic patient.

   a. 1 and 2 only
   b. 3 and 4 only
   c. 1, 2, and 3 only
   *d. 1, 2, 3, and 4

6. An AP shoulder radiograph taken on a patient whose upper midcoronal plane was tilted anteriorly demonstrates:
   1. the acromion inferior to the coracoid.
   2. the scapular body longitudinally foreshortened.
   3. the superior scapular angle superior to the clavicle.
   4. the thoracic cavity with increased scapular body superimposition.

   a. 1 and 4 only
   b. 2 and 3 only
   c. 2, 3, and 4 only
   *d. 1, 2, 3, and 4

7. How can the positioning setup be adjusted for an AP shoulder radiograph to demonstrate uniform density throughout the shoulder and clavicular areas?
   1. Position the top of the shoulder at the cathode end of the tube.
   2. Place a compensating filter over or under the laterally located acromion and clavicular end.
   3. Use a kilovoltage above 80.
   4. Use a grid.

   a. 1 and 2 only
   *b. 2 only
   c. 1 and 3 only
   d. 2, 3, and 4 only

8. An accurately positioned axial shoulder radiograph demonstrates:
   1. the glenoid fossa without humeral head superimposition.
   2. nearly superimposed inferior and superior glenoid fossa margins.
   3. the lesser tubercle in profile anteriorly.
   4. the lateral edge of the coracoid base medial to the glenoid fossa.

   a. 1 and 2 only
   b. 3 and 4 only
   *c. 1, 2, and 3 only
   d. 1, 2, 3, and 4

9. For an axial shoulder radiograph:
   1. the patient's shoulder is elevated on a sponge or washcloth.
   2. the patient's head is tilted and rotated toward the affected shoulder.
   3. the patient's affected arm is externally rotated.
   4. a 30 to 35 degree central ray to lateral body surface angle is used if the arm is adequately abducted.

   a. 1 and 2 only
   b. 2 and 3 only
   *c. 1, 3, and 4 only
   d. 1, 2, 3, and 4

10. An axial shoulder radiograph taken with too large of a central ray to lateral body angle demonstrates:
    1. the glenoid fossa projected medial to the lateral edge of the coracoid base.
    2. the glenoid fossa projected lateral to the coracoid base.
    3. the lesser tubercle in profile anteriorly.
    4. the posterior surface of the acromion and humerus on the radiograph.

    *a. 1 and 3 only
    b. 2 and 4 only
    c. 2, 3, and 4 only
    d. 1, 3, and 4 only

11. An axial shoulder radiograph taken with the humerus in exaggerated external rotation demonstrates:
    1. the Hill-Sachs defect.
    2. the humeral neck in profile.
    3. the lesser tubercle in profile anteriorly.
    4. the greater trochanter in profile posteriorly.

    *a. 1 and 3 only
    b. 2 and 4 only
    c. 1, 3, and 4 only
    d. 1, 2, 3, and 4

12. An axial shoulder radiograph that does not include the posterior aspects of the acromion and humerus:
    a. was taken without the patient's head rotated or tilted toward the unaffected shoulder.
    b. was taken using too large of a central ray to lateral body surface angle.
    c. was taken without the patient's shoulder elevated on a sponge or washcloth.
    d. was taken without the film positioned proximally enough.

13. When is it necessary to use a grid for an axial shoulder radiograph?
    1. The <u>anteroposterior</u> measurement is over 5 inches (13 cm).
    2. The kilovoltage used is above 70.
    3. The <u>inferosuperior</u> measurement is over 5 inches (13 cm).
    4. The kilovoltage used is below 70.

    a. 1 and 2 only
    *b. 2 and 3 only
    c. 1 and 4 only
    d. 3 and 4 only

14. For a Grashey method radiograph:
    1. the patient's midsagittal plane is rotated to a 45-degree angle with the film.
    2. the central ray is centered 0.75 inch (2 cm) inferior to the coracoid.
    3. the patient is rotated toward the affected shoulder.
    4. the radiograph is taken with the patient in an upright position.

    a. 1 and 3 only
    b. 2 and 4 only
    *c. 2, 3, and 4 only
    d. 1, 2, 3, and 4

15. For a Grashey shoulder radiograph the patient is rotated more than 45 degrees when:
    1. the patient is recumbent.
    2. the patient is kyphotic.
    3. the patient is seated.
    4. the patient is upright and leaning against the upright film holder.

    a. 1 and 2 only
    b. 3 and 4 only
    *c. 1, 2, and 4 only
    d. 1, 2, 3, and 4

16. An accurately positioned Grashey shoulder radiograph demonstrates:
    1. the glenoid fossa in profile and facing superiorly.
    2. an open glenohumeral joint space.
    3. a transversely foreshortened clavicle.
    4. the glenohumeral joint in the center of the collimated field.

    a. 1 and 3 only
    b. 1, 2, and 4 only
    *c. 2, 3, and 4 only
    d. 1, 2, 3, and 4

17. A Grashey shoulder radiograph taken with the patient rotated less than required to obtain accurate positioning, demonstrates:
    1. more than just the lateral tip of the coracoid superimposing the humeral head.
    2. a closed glenohumeral joint space.
    3. excessive transverse clavicular foreshortening.
    4. an increase in the amount of thorax and scapular body superimposition.

    *a. 2 only
    b. 1, 2, and 3 only
    c. 3 and 4 only
    d. 1, 2, 3, and 4

18. The arms of the Y on a transcapular Y shoulder radiograph are formed by the:
    1. coracoid
    2. scapular body
    3. acromion
    4. glenoid fossa

    a. 1 and 2 only
    b. 2 and 3 only
    c. 3 and 4 only
    *d. 1 and 3 only

19. For a transcapular Y shoulder radiograph:
    1. the patient's humerus is elevated until the hand is placed on the hip.
    2. the patient is rotated toward the affected shoulder.
    3. the patient is rotated until an imaginary line drawn connecting the acromion angle and coracoid is aligned parallel with the film.
    4. the patient's midcoronal plane is vertical.

    a. 2 and 3 only
    b. 2 and 4 only
    *c. 2, 3, and 4 only
    d. 1, 2, 3, and 4

20. An accurately positioned transcapular Y shoulder radiograph demonstrates:
    1. the superior angle of the scapula superimposing the clavicle.
    2. superimposed scapular borders.
    3. a laterally situated glenoid fossa.
    4. the coracoid, acromion, and glenoid fossa create the arms and leg of the Y formation.

    *a. 1 and 2 only
    b. 2 and 4 only
    c. 1 and 3 only
    d. 1, 2, and 4 only

21. An accurately positioned transcapular Y shoulder radiograph on a patient with an anterior dislocation demonstrates:
    1. a Y formation.
    2. the humeral head positioned anterior to the glenoid fossa beneath the coracoid.
    3. the humeral head positioned anterior to the glenoid fossa beneath the acromion.
    4. the humerus superimposing the scapular body.

    a. 1 and 3 only
    *b. 1 and 2 only
    c. 1, 2, and 4 only
    d. 2 and 4 only

22. A transcapular Y shoulder radiograph taken with the patient in a posterior oblique demonstrates:
    1. the humeral head superimposing the glenoid fossa.
    2. the glenoid fossa on-end.
    3. the medial scapular border slightly medial to the lateral border.
    4. magnification of the scapula and humerus.

    a. 1 and 3 only
    b. 2 and 4 only
    *c. 1, 2, and 4 only
    d. 1, 2, 3, and 4

23. A transcapular Y shoulder radiograph that was taken with the patient over rotated demonstrates:
    1. the glenoid fossa in profile medially.
    2. the medial scapular border closer to the ribs than the lateral scapular border.
    3. the glenoid fossa in profile laterally.
    4. the lateral scapular border closer to the ribs than the medial scapular border.

    a. 1 and 2 only
    *b. 1 and 4 only
    c. 2 and 3 only
    d. 3 and 4 only

24. A transcapular Y shoulder radiograph taken with the patient's upper midcoronal plane tilted toward the film demonstrates:
    1. a Y formation.
    2. the superior scapular angle superior to the coracoid.
    3. a longitudinally foreshortened scapula.
    4. the glenoid fossa medially.

    a. 1 and 4 only
    b. 2 and 3 only
    *c. 1, 2, and 3 only
    d. 1, 2, 3, and 4

25. An accurately positioned AP clavicular radiograph demonstrates:
    1. the medial clavicular end next to the lateral edge of the vertebral column.
    2. the superior scapular angle superior to the clavicle.
    3. inferosuperior foreshortening on the kyphotic patient unless the central ray is angled cephalically.
    4. an overexposed medial clavicle unless a compensating filter is used.

    a. 1 and 2 only
    *b. 1 and 3 only
    c. 3 and 4 only
    d. 1, 2, and 3 only

26. An AP clavicular radiograph taken with the patient rotated away from the affected shoulder demonstrates:
    1. the medial clavicular end superimposing the vertebral column.
    2. the medial clavicular end shifted away from the vertebral column.
    3. the scapular body with increased thorax superimposition.
    4. the scapular body with decreased thorax superimposition.

    a. 1 and 3 only
    *b. 1 and 4 only
    c. 2 and 3 only
    d. 2 and 4 only

27. An accurately positioned AP axial clavicular radiograph demonstrates:
    1. the medial clavicular end superimposing the first and second ribs.
    2. the middle and lateral thirds of the clavicle superior to the acromion.
    3. the clavicle bowing upwardly.
    4. the medial clavicular end superimposing the vertebral column.

    a. 1 and 2 only
    b. 2 and 3 only
    *c. 1, 2, and 3 only
    d. 1, 2, 3, and 4

28. For an AP axial clavicular radiograph:
    1. the patient's shoulders is positioned at equal distances from the film.
    2. the central ray is angled 15 to 30 degrees cephalad.
    3. a compensating filter is positioned over or under the lateral clavicle.
    4. the central ray is centered halfway between the medial and lateral clavicular ends.

    a. 2 and 4 only
    *b. 1 and 3 only
    c. 1, 3, and 4 only
    d. 1, 2, 3, and 4

29. For an AP without-weights AC joint radiograph:
   1. the central ray is centered to the AC joint.
   2. the midcoronal plane is positioned parallel with the film.
   3. a grid is used if the patient measures 6 inches (15 cm).
   4. the patient's shoulders are positioned at equal distances from the film.

   a. 1 and 4 only
   b. 3 only
   c. 1 and 2 only
   *d. 1, 2, 3, and 4

30. An accurately positioned AP scapular radiograph demonstrates:
   1. nearly superimposed anterior and posterior glenoid fossa margins.
   2. the medial scapular border without thoracic cavity superimposition.
   3. the humeral shaft at a 90-degree angle with the body.
   4. the scapular body with slight longitudinal foreshortening.

   *a. 1 and 3 only
   b. 1, 3, and 4 only
   c. 1 and 2 only
   d. 3 and 4 only

31. For an AP scapular radiograph:
   1. the patient's arm is abducted 90 degrees to the body.
   2. the radiograph is taken on expiration.
   3. the upper midcoronal plane is lean slightly away from the film.
   4. the central ray is centered 3 inches (7.5 cm) inferior to the coracoid.

   a. 1 and 4 only
   b. 2 and 3 only
   c. 1, 3, and 4 only
   *d. 1, 2, and 4 only

32. An accurately positioned lateral scapular radiograph taken with the humerus abducted to a 90-degree angle with the body demonstrates:
   1. the lateral and vertebral scapular borders are superimposed.
   2. the scapula in a Y formation.
   3. the humerus superimposing the scapular body.
   4. the mid scapular body in the center of the collimated field.

   *a. 1 and 4 only
   b. 2 and 4 only
   c. 1, 2, and 4 only
   d. 1, 2, and 3 only

33. A lateral scapular radiograph taken with the patient inadequately rotated and the arm placed at a 90-degree angle with the patient demonstrates:
   1. superimposed lateral and vertebral scapular borders.
   2. the lateral scapular border medial to the vertebral border.
   3. the superior scapular angle inferior to the coracoid.
   4. the vertebral scapular border medial to the lateral border.

   a. 1 and 3 only
   b. 2 only
   c. 2 and 3 only
   *d. 4 only

34. An imaginary line connecting the humeral epicondyles is positioned perpendicular to the film for:
    1. an internally rotated AP shoulder radiograph.
    2. a Grashey method shoulder radiograph.
    3. a lateral humeral radiograph.
    4. a transcapular Y shoulder radiograph.

    *a. 1 and 3 only
    b. 2 and 4 only
    c. 3 only
    d. 1, 3, and 4 only

35. The lesser tubercle is demonstrated in profile for:
    1. a neutral AP shoulder radiograph.
    2. a lateral humeral radiograph.
    3. a transthoracic humeral or shoulder radiograph.
    4. an axial lateral shoulder radiograph.

    a. 1 and 3 only
    b. 2 and 4 only
    c. 1, 3, and 4 only
    *d. 2, 3, and 4 only

36. The glenohumeral joint space is visualized as an open space on:
    1. an axial lateral shoulder radiograph.
    2. a transthoracic lateral shoulder radiograph.
    3. a Grashey method shoulder radiograph.
    4. a transcapular Y shoulder radiograph.

    a. 1 only
    *b. 1 and 3 only
    c. 2 and 3 only
    d. 1, 2, 3, and 4

Chapter 5

Radiographic Critique of the Lower Extremity

I. Lecture on radiographic projections/positions of the lower extremity.

II. Lecture on radiographic evaluation for each projection/position of the lower extremity.

III. Classroom activities.

    A. Workbook study questions.

    B. Workbook learning activities numbers 2, 3, and 4.

Suggestion for learning activity number 2: I provide several connected, movable lower extremity bones, and accurately and poorly positioned lower extremity radiographs. The students then work together, identifying the anatomical structures on the bones and on the radiographs.

Suggestion for learning activity number 3: See chapter 2, suggestions for learning activity number 4, page 7. Fill the packets with poorly positioned lower extremity radiographs.

Suggestion for learning activity number 4: I ask the students to each bring radiographs from the lower extremity repeat bin to class. During class they label all the anatomical structures found on the radiographs.

IV. Clinical laboratory activities.

    A. Workbook learning activity number 1.

        Suggestion for learning activity number 1: I assign each student a lower extremity projection/position to demonstrate to the class. See chapter 2, suggestions for learning activity number 1, page 7 for further description.

V. Workbook self test questions.

    Suggestion: I ask the students not to look at or do the self test prior to completing the study question and learning activities. After the study questions and learning activities are completed the self test is taken in the classroom setting as if it were a quiz.

    Used in this manner the self test is a tool that pinpoints areas of weakness before the final exam.

VI. Final exam.

    The following questions pertain to both radiographic positioning and critique, and can be incorporated in your present positioning and critiquing exams. They are based on information found in the Radiographic Critique book. Since some facilities do not follow all the protocols found in the Radiographic Critique book, evaluate each question for information that does not match your facility's routines.

1. The IP and MP joint spaces on a toe radiograph are open when:
   1. the central ray is aligned parallel with them.
   2. the central ray is aligned perpendicular to them.
   3. the joints are aligned parallel with the film.
   4. the joints are aligned perpendicular to the film.

     a. 1 and 3 only
   *b. 1 and 4 only
     c. 2 and 3 only
     d. 2 and 4 only

2. For an oblique second toe radiograph the toe is:
     a. rotated 30 degrees laterally.
     b. rotated 45 degrees laterally.
     c. rotated 30 degrees medially.
   *d. rotated 45 degrees medially.

3. An AP first toe radiograph that was taken with the foot and toe rotated 45 degrees medially demonstrates:
   1. equal soft tissue width on each side of the phalanges.
   2. more midshaft concavity on one side of the phalanges than the opposite side.
   3. twice as much soft tissue on one side of the phalanges than the opposite side.
   4. convexity on one side on the phalanges and concavity on the opposite side.

     a. 1 and 2 only
     b. 1 and 4 only
   *c. 2 and 3 only
     d. 3 and 4 only

4. For a lateral fourth toe radiograph:
   1. the foot is rotated laterally until the toe is in a lateral position.
   2. the adjacent toes are drawn away from the affected toe.
   3. the long axis of the digit is aligned with the transverse axis of the collimated field.
   4. the central ray is centered to the IP joint.

     a. 1 and 2 only
     b. 2 and 3 only
   *c. 1, 2, and 4 only
     d. 1, 3, and 4 only

5. Which of the following positioning setup procedures must be completed to obtain open tarsometatarsal and navicular-cuneiform joint spaces on an AP foot radiograph?
   1. The patient's foot is positioned flat against the film.
   2. The foot, ankle, and lower leg are aligned.
   3. The central ray is angled 10 to 15 degrees proximally.
   4. A compensating filter is placed over the toes.

   *a. 1 and 3 only
   b. 3 only
   c. 1, 2, and 3 only
   d. 1 and 4 only

6. Where should the central ray enter for a dorsoplantar projection of the foot?
   a. 3rd metatarsophalangeal joint
   *b. Base of the 3rd metatarsal
   c. Anterior talus
   d. Intermediate cuneiform

7. An AP foot radiograph taken with the foot laterally rotated demonstrates:
   1. a closed medial-intermediate cuneiform joint space.
   2. closed tarsometatarsal joint spaces.
   3. the calcaneus with increased talar superimposition.
   4. a decrease in metatarsal base superimposition.

   *a. 1 and 3 only
   b. 1 and 4 only
   c. 2 and 3 only
   d. 1, 2, and 4 only

8. How can the positioning setup procedure be adjusted for an AP foot radiograph to demonstrate uniform radiographic density throughout the toes and foot areas?
   1. Position the toes at the cathode end of the tube.
   2. Use a kVp above 75.
   3. Use a grid.
   4. Place a compensating filter over the toes.

   a. 1 only
   b. 2 and 3 only
   c. 1 and 4 only
   *d. 4 only

9. An accurately positioned oblique foot radiograph demonstrates:
   1. open first and second intermetatarsal joint spaces.
   2. open joint spaces around the cuboid.
   3. slight superimposition of the fourth and fifth metatarsal bases.
   4. the long axis of the foot aligned with the long axis of the collimated field.

   a. 1 and 2 only
   b. 1 and 3 only
   c. 1, 3, and 4 only
   *d. 2 and 4 only

10. What joint spaces are open on an accurately positioned oblique foot radiograph?
    1. The third through fifth intermetatarsal joints.
    2. The navicular-cuneiform.
    3. The joint spaces surrounding the cuboid.
    4. The tarsometatarsal.

    *a. 1 and 3 only
    b. 3 only
    c. 1, 3, and 4 only
    d. 1, 2, 3, and 4

11. Which of the following pertains to the positioning setup for an oblique foot radiograph on a patient with a high longitudinal arch?
    1. Oblique the patient's foot 45 degrees.
    2. Angle the central ray 15 degrees proximally.
    3. Align the long axis of the foot with the long axis of the collimated field.
    4. Center the central ray to the third metatarsal base.

    a. 1 and 3 only
    b. 2 and 4 only
    *c. 1, 3, and 4 only
    d. 1, 2, 3, and 4

12. On an accurately positioned lateral foot radiograph:
    1. the medial talar dome should be demonstrated slightly superior to the lateral dome.
    2. tibiotalar joint space should be open.
    3. the talar domes are superimposed.
    4. the distal fibula is superimposed by the posterior half of the distal tibia.

    a. 1 and 3 only
    b. 2 and 4 only
    *c. 2, 3, and 4 only
    d. 3 and 4 only

13. What aspect of the foot is placed parallel with the film for the routine lateral foot position?

    a. Plantar
    b. Dorsal
    *c. Lateral
    d. Medial

14. If the medial talar dome was positioned distal to the lateral talar dome on a lateral foot radiograph, which of the following are true?
    a. The patient's heel was elevated off the cassette.
    *b. The patient's proximal tibia was elevated.
    c. The patient's forefoot and toes were elevated off the cassette.
    d. The patient's distal lower leg was elevated.

15. A lateral foot radiograph taken on a patient whose leg was rotated anteriorly (heel elevated off film) demonstrates:
    1. more than 0.5 inch (1 cm) of the cuboid distal to the navicular.
    2. the fibula situated too posterior on the tibia.
    3. the lateral talar dome anterior to the medial talar dome.
    4. an obscured tibiotalar joint space.

    a. 1 and 2 only
    b. 2 only
    c. 3 only
    *d. 2 and 4 only

16. Which of the following pertains to a lateral foot radiograph that demonstrates the lateral talar dome distal to the medial talar dome?
    1. The patient was radiographed with the distal tibia elevated.
    2. More than 0.5 inch (1 cm) of the cuboid is demonstrated distal to the navicular.
    3. The lateral talar dome is also anterior to the medial talar dome.
    4. The fibula would be situated too far posterior on the tibia.

    a. 1 only
    *b. 1 and 2 only
    c. 3 and 4 only
    d. 1, 2, 3, and 4

17. Which of the following is true about axial calcaneal radiography?
    1. The radiograph demonstrates an open talocalcaneal joint space.
    2. The foot is flexed 90 degrees to the lower leg and rotated slightly laterally.
    3. A 40-degree central ray is directed proximally.
    4. The central ray is centered to the distal fifth metatarsal.

    *a. 1 and 3 only
    b. 1 and 2 only
    c. 3 and 4 only
    d. 1, 3, and 4 only

18. An axial calcaneal radiograph taken with the patient's foot in plantoflexion and the central ray angled 40 degrees proximally demonstrates.
    1. an elongated calcaneal tuberosity.
    2. a foreshortened calcaneal tuberosity.
    3. an open talocalcaneal joint space.
    4. a closed talocalcaneal joint space.

    a. 1 and 3 only
    b. 1 and 4 only
    c. 2 and 3 only
    *d. 2 and 4 only

19. For an AP ankle radiograph:
    1. the intermalleolar line is aligned at a 15 to 20 degree angle with the film.
    2. the lateral malleolus is positioned posterior to the medial malleolus.
    3. the long axis of the foot is positioned perpendicular with the film.
    4. the central ray is centered at the level of the medial malleolus.

    a. 2 and 4 only
    b. 2, 3, and 4 only
    c. 1, 3, and 4 only
    *d. 1, 2, 3, and 4

20. An AP ankle radiograph taken with the patient's leg in lateral rotation will demonstrate which of the following?
    1. A closed medial mortise.
    2. Decreased talar and fibular superimposition.
    3. An open lateral mortise.
    4. The sinus tarsi.

    *a. 1 only
    b. 1 and 2 only
    c. 2, 3, and 4 only
    d. 1, 2, 3, and 4

21. An accurately obliqued ankle radiograph demonstrates which of the following joints as open spaces?
    1. Tibiotalar
    2. Talofibular
    3. Lateral mortise
    4. Medial mortise

    a. 1 and 4 only
    b. 2 and 3 only
    c. 1, 2, and 4 only
    *d. 1, 2, and 3 only

22. For an oblique ankle radiograph:
    1. the central ray is centered at the level of the medial malleolus.
    2. the foot is dorsiflexed to a 90-degree angle with the lower leg.
    3. the long axis of the lower leg is aligned with the long axis of the collimated field.
    4. the leg is internally rotated until the intermalleolar line is parallel with the film.

    a. 1 and 2 only
    b. 2 and 4 only
    c. 1 and 3 only
    *d. 1, 2, 3, and 4

23. A poorly positioned oblique ankle radiograph demonstrates an open distal lateral mortise superimposing the calcaneus. How was the patient mispositioned to obtain such a radiograph?
    *a. The foot was plantar flexed.
    b. The leg was not adequately internally rotated.
    c. The central ray was centered too caudally.
    d. The proximal lower leg was elevated.

24. Which of the following are reasons why the foot should be dorsiflexed to a 90-degree angle with the lower leg for a lateral foot radiograph.
    1. It places the tibiotalar joint in a neutral position.
    2. It prevents the patient from rotating posteriorly.
    3. To allow the anterior pretalar fat pad to be used to detect joint effusion.
    4. It positions the talar domes on top of each other.

    *a. 1 and 3 only
    b. 1, 2, and 3 only
    c. 3 only
    d. 1 and 4 only

25. For a lateral ankle radiograph:
    1. the medial and lateral malleoli are positioned on top of each other.
    2. the lateral foot surface is aligned parallel with the film.
    3. the lower leg is parallel with the tabletop.
    4. the central ray is centered to the medial malleolus.

    a. 2 and 4 only
    b. 1 and 3 only
    *c. 2, 3, and 4 only
    d. 1, 2, 3, and 4

26. An accurately positioned lateral ankle radiograph demonstrates:
    1. an open tibiotalar joint.
    2. about 0.5 inch (1 cm) of the cuboid distal to the navicular.
    3. 1 inch (2.5 cm) of the fifth metatarsal base.
    4. the fibula in the posterior half of the tibia.

    a. 1, 2, and 3 only
    b. 2 and 4 only
    c. 1, 3, and 4 only
    *d. 1, 2, 3, and 4

27. A lateral ankle radiograph demonstrates the fibula too anterior on the tibia and less than 0.5 inch (1 cm) of the cuboid distal to the navicular. How are the talar domes positioned on this radiograph?
    1. Medial dome anterior.
    2. Medial dome proximal.
    3. Lateral dome anterior.
    4. Lateral dome proximal.

    a. 1 and 2 only
    b. 2 and 3 only
    *c. 3 and 4 only
    d. 1 and 4 only

28. For an accurately positioned AP projection of the knee:
    1. an imaginary line connecting the femoral epicondyles is aligned parallel with the film.
    2. the intercondylar eminence is centered within the intercondylar fossa.
    3. the long axis of the foot is tilted internally.
    4. the femoral condyles are symmetrical.

    a. 1 and 2 only
    b. 2 and 4 only
    c. 1, 2 and 4 only
    *d. 1, 2, 3 and 4

29. A cephalic central ray angulation is required on an AP knee radiograph when:
    1. the exam is taken in an upright position.
    2. the patient's anterior tibial margin is distal to the posterior tibial margin.
    3. the patient's ASIS to table-top measurement is 22 centimeters.
    4. the knee is flexed and a curved cassette is used.

    a. 2 and 3 only
    b. 4 only
    *c. 2 and 4 only
    d. 1, 2 and 4 only

30. If the patient is unable to fully extend the knee, an open femorotibial joint is accomplished by aligning the central ray perpendicular to the lower leg:
    *a. then decreasing the angle 3 - 5 degrees and centering to the femorotibial joint.
    b. then increasing the angle 3 - 5 degrees and centering to the femorotibial joint.
    c. then centering to the femorotibial joint.

31. The position of the patella on an AP knee radiograph is affected by:
    1. patellar subluxation.
    2. knee rotation.
    3. knee flexion.
    4. foot inversion

    a. 2 only
    b. 1 and 3 only
    *c. 1, 2, and 3 only
    d. 1, 2, 3, and 4

32. An AP knee radiograph taken with the central ray angled too cephalically demonstrates:
    1. the proximal ridges of the femoral condyle with a concave contour.
    2. a foreshortened fibular head.
    3. the fibular head at a position less than 0.5 inch (1 cm) distal to the tibial plateau.
    4. a narrowed or closed femorotibial joint space.

    a. 1, 3, and 4 only
    *b. 2 and 4 only
    c. 1, 2, and 3 only
    d. 1, 2, 3, and 4

33. An AP knee radiograph taken with the knee medially rotated demonstrates:
    1. The larger appearing medial femoral condyle than lateral condyle.
    2. The lateral femoral condyle will appear larger than the medial condyle.
    3. The fibular head with increased tibial superimposition.
    4. The fibular head with decreased tibial superimposition.

    a. 1 and 3 only
    b. 1 and 4 only
    c. 2 and 3 only
    *d. 2 and 4 only

34. For an accurately positioned laterally obliqued knee radiograph:
    1. the fibular head is demonstrated free of tibial superimposition.
    2. the lateral femoral condyle is visualized in profile.
    3. the fibular head, neck, and shaft are superimposed by the tibia.
    4. the medial condyle is visualized in profile.

    a. 1 and 2 only
    b. 1 and 4 only
    c. 2 and 3 only
    *d. 3 and 4 only

35. For a lateral oblique knee radiograph:
    1. the leg is externally rotated until an imaginary line connecting the femoral epicondyles is at a 45-degree angle with the film.
    2. the leg is internally rotated until an imaginary line connecting the femoral epicondyles is at a 45-degree angle with the film.
    3. the central ray is aligned parallel with the tibial plateau.
    4. the central ray is centered at a level 0.75 inch (2 cm) distal to the medial femoral epicondyle.

    a. 1 and 4 only
    b. 1 and 3 only
    c. 2 and 4 only
    *d. 1, 3, and 4 only

36. For a lateral knee radiograph:
    1. an imaginary line connecting the femoral epicondyles is aligned parallel with the film.
    2. a patient with long femora and a narrow pelvis does not require an angled central ray.
    3. a grid is used if the knee measures 10 cm.
    4. the central ray is centered 0.75 inch (2 cm) distal to the medial femoral epicondyle.

    a. 1 and 2 only
    *b. 2 and 4 only
    c. 3 and 4 only
    d. 1, 2, 3, and 4

37. An accurately positioned lateral knee radiograph demonstrates:
    1. superimposed femoral condyles.
    2. the fibular head without tibial superimposition.
    3. an open femorotibial joint space.
    4. one-fourth of the distal femur and proximal lower leg.

    a. 1 and 3 only
    b. 2 and 4 only
    *c. 1, 3, and 4 only
    d. 1, 2, 3, and 4

38. If the medial femoral condyle is situated anterior to the lateral femoral condyle on a poorly positioned lateral knee radiograph which of the following is true?
    1. The fibular head demonstrates increased tibial superimposition.
    2. The adductor tubercle will be located on the anterior condyle.
    3. The distal surface of the anterior condyle will appear flatter.
    4. The fibular head will demonstrate a decrease in tibial superimposition.

    a. 1 and 2 only
    b. 2 and 3 only
    *c. 2 and 4 only
    d. 2, 3, and 4 only

39. A lateral knee radiograph demonstrates the medial femoral condyle anterior and proximal to the medial femoral condyle. How was the positioning setup mispositioned to obtain this radiograph?
    1. The central ray was angled too caudally.
    2. The central ray was angled too cephalically.
    3. The patient's patella was positioned too close to the film.
    4. The patient's patella was positioned too far away from the film.

    a. 1 and 3 only
    b. 1 and 4 only
    *c. 2 and 3 only
    d. 2 and 4 only

40. If the patient's patella was positioned too close to the film for a lateral knee radiograph:
    1. the fibula would be demonstrated with increased tibial superimposition.
    2. the fibula would be demonstrated with decreased tibial superimposition.
    3. the medial condyle would be demonstrated anteriorly to the lateral condyle.
    4. the lateral condyle would be demonstrated anteriorly to the medial condyle.

    a. 1 and 3 only
    *b. 2 and 3 only
    c. 1 and 4 only
    d. 2 and 4 only

41. A 5 to 7 degree central ray angulation is used for a lateral knee radiograph:
    1. to project the medial condyle anterosuperiorly.
    2. on a patient with a narrow pelvis and long femora.
    3. to off-set the reduction in medial inclination that occurs when the patient is in a lateral recumbent position.
    4. to obtain an open femorotibial joint space.

    a. 1 and 2 only
    b. 2 and 3 only
    *c. 1, 3, and 4 only
    d. 2, 3, and 4 only

42. Positioning the femur at a 60 to 70 degree angle with the tabletop for the Holmblad position:
    1. superimposes the proximal surfaces of the intercondylar fossa.
    2. places the patellar apex superior to the intercondylar fossa.
    3. superimposes the lateral and the medial surfaces of the intercondylar fossa.
    4. superimposes the anterior and posterior margins of the tibial plateau.

    a. 1 only
    *b. 1 and 2 only
    c. 3 and 4 only
    d. 1, 2 and 3 only

43. Proper elevation of the distal lower leg and vertical placement of the foot's long axis (heel is not rotated side to side) for the Holmblad position:
    1. superimposes the proximal surfaces of the intercondylar fossa.
    2. places the patellar apex superior to the intercondylar fossa.
    3. superimposes the lateral and the medial surfaces of the intercondylar fossa.
    4. superimposes the anterior and posterior margins of the tibial plateau.

    a. 1 and 2 only
    b. 4 only
    *c. 3 and 4 only
    d. 2, 3 and 4 only

44. If a Holmblad knee radiograph is taken with the patient's heel rotated internally which of the following is true?
    1. The proximal surfaces of the intercondylar fossa are not superimposed.
    2. The lateral and the medial surfaces of the intercondylar fossa are not superimposed.
    3. The patella is rotated laterally.
    4. The tibia is demonstrated without fibular head superimposition.

    a. 1 and 3 only
    b. 2 and 3 only
    *c. 2, 3, and 4 only
    d. 1 and 4 only

45. For a Merchant knee radiograph:
    1. an imaginary line connecting the femoral epicondyles is aligned parallel with the tabletop.
    2. the medial condyles demonstrate more height than the lateral condyles.
    3. the femorotibial joints are open.
    4. the patient is instructed to relax the leg muscles.

    *a. 1 and 4 only
    b. 1 and 2 only
    c. 2 and 3 only
    d. 1, 2, 3, and 4

46. If the curves of the posterior knees are not accurately positioned just above the bend of the "axial viewer":
    1. the patellae are projected into the patellofemoral joint spaces.
    2. the tibial tuberosities are demonstrated within the joint spaces.
    3. the soft tissue from the anterior thighs are projected into the joint spaces.
    4. the knees are flexed more or less than 45 degrees.

    a. 1 and 2 only
    b. 3 only
    *c. 1, 2, and 4 only
    d. 3 and 4 only

47. The Merchant position is a _____ projection.
    a. inferosuperior
    b. mediolateral
    *c. superoinferior
    d. anteroposterior

48. When the legs are flexed 30 degrees for the Merchant position the central ray should be angled:
    *a. 75 degrees
    b. 45 degrees
    c. 60 degrees
    d. perpendicular to the lower leg

Chapter 6
Radiographic Critique of the Hip and Pelvis

I. Lecture on radiographic projections/positions of the hip and pelvis.

II. Lecture on radiographic evaluation for each projection/position of the hip and pelvis.

III. Classroom activities.

   A. Workbook study questions.

   B. Workbook learning activities numbers 2, 3, and 4.

   Suggestion for learning activity number 2: I provide several connected and unconnected femoral and pelvic bones, and accurately and poorly positioned hip and pelvic radiographs. The students then work together, identifying the anatomical structures on the bones and on the radiographs.

   Suggestion for learning activity number 3: See chapter 2, suggestion for learning activity number 4, page 7. Fill the packets with poorly positioned hip and pelvic radiographs.

   Suggestion for learning activity number 4: I ask the students to each bring radiographs from the hip and pelvis repeat bin to class. During class they label all the anatomical structures found on the radiographs.

IV. Clinical laboratory activities.

   A. Workbook learning activity number 1.

   Suggestion for learning activity number 1: I assign each student a hip and pelvis projection/position to demonstrate to the class. See chapter 2, suggestions for learning activity number 1, page 7 for further description.

V. Workbook self test questions.

   Suggestion: I ask the students not to look at or do the self test prior to completing the study question and learning activities. After the study questions and learning activities are completed the self test is taken in the classroom setting as if it were a quiz.
   Used in this manner the self test is a tool that pinpoints areas of weakness before the final exam.

VI. Final exam.
   The following questions pertain to both radiographic positioning and critique, and can be incorporated in your present positioning and critiquing exams. They are based on information found in the Radiographic Critique book. Since some facilities do not follow all the protocols found in the Radiographic Critique book, evaluate each question for information that does not match your facility's routines.

1. For an accurately positioned AP projection of the hip:
   1. the ASISs are positioned at equal distances to the film.
   2. the patient's legs are externally rotated until the epicondyles are at a 45-degree angle with the tabletop.
   3. gonadal shielding should not be used.
   4. the central ray is centered 1.5 inches (4 cm) medial to the ASIS at the symphysis pubis.

   *a. 1 and 4 only
   b. 2 and 3 only
   c. 1, 3, and 4 only
   d. 2 and 3 only

2. A left AP hip radiograph taken on a patient that was rotated toward the right side demonstrates:
   1. a narrowed left obturator foramen.
   2. the sacrum and coccyx rotated toward the left hip.
   3. a narrowed left iliac wing.
   4. the lesser trochanter in profile.

   a. 1 only
   *b. 2 and 3 only
   c. 1, 2, and 3 only
   d. 2, 3, and 4 only

3. The patient's leg is internally rotated for an AP hip radiograph:
   1. to bring the lesser trochanter in profile.
   2. to decrease femoral neck foreshortening.
   3. to bring the greater trochanter in profile.
   4. to increase femoral neck foreshortening.

   a. 1 and 2 only
   *b. 2 and 3 only
   c. 3 and 4 only
   d. 1, 3 and 4 only

4. A nonrotated AP pelvis radiograph demonstrates:
   1. the sacrum and coccyx aligned with the symphysis pubis.
   2. the ischial spines without pelvic brim superimposition.
   3. a narrow right iliac wing and a wider left iliac wing.
   4. symmetrically appearing obturator foramina.

   a. 1 only
   *b. 1 and 4 only
   c. 1, 2 and 4 only
   d. 3 and 4 only

5. An AP pelvis radiograph taken with the patient rotated toward the left hip (LPO) demonstrates:
   1. the symphysis pubis rotated toward the left hip.
   2. a narrower right iliac wing.
   3. a narrower left obturator foramen.
   4. the sacrum and coccyx rotated toward the right hip.

   a. 1 and 4 only
   b. 2 and 3 only
   c. 3 and 4 only
   *d. 1, 2, 3, and 4

6. An AP hip radiograph taken with the patient's leg in external rotation demonstrates:
   1. the lesser trochanter in profile.
   2. a foreshortened femoral neck.
   3. the greater trochanter in profile.
   4. the femoral neck without foreshortening.

   *a. 1 and 2 only
   b. 1 and 4 only
   c. 2 and 3 only
   d. 3 and 4 only

7. For a frogleg hip radiograph the patient was positioned with the left ASIS placed closer to the tabletop than the right ASIS. On such a radiograph the left hip demonstrates:
   1. a narrowed obturator foramen
   2. a widened iliac wing.
   3. the iliac spine without pelvic brim superimposition.
   4. the sacrum and coccyx without symphysis pubis alignment.

   a. 1 and 2 only
   b. 2, 3, and 4 only
   c. 1, 3, and 4 only
   *d. 1, 2, 3, and 4.

8. For a frogleg hip radiograph:
   1. the lesser trochanter is demonstrated in profile.
   2. the greater trochanter at the same transverse level as the femoral head.
   3. the ischial spine with pelvic brim superimposition.
   4. the greater trochanter is demonstrated medially.

   a. 1 and 2 only
   *b. 2 and 3 only
   c. 1, 2, and 3 only
   d. 1, 2, 3, and 4

9. A frogleg hip radiograph taken with the leg abducted nearly to the tabletop demonstrates:
   1. the greater trochanter is at a transverse level halfway between the lesser trochanter and femoral head.
   2. the greater trochanter laterally.
   3. the greater trochanter superimposed by the femoral head.
   4. the greater trochanter medially.

   a. 1 and 2 only
   *b. 3 only
   c. 3 and 4 only
   d. 4 only

10. A frogleg hip radiograph taken with the knee overflexed demonstrates:
    1. an obscured lesser trochanter.
    2. the greater trochanter laterally.
    3. the greater trochanter superimposing the femoral head.
    4. the greater trochanter medially.

    *a. 1 and 4 only
    b. 1 and 2 only
    c. 2 only
    d. 3 only

11. For a frogleg pelvis radiograph:
    1. the legs are abducted until the femurs are at a 60 to 70 degree angle with tabletop.
    2. the ASISs are positioned at equal distances to the tabletop.
    3. the knees and hips are flexed until the femurs are aligned at a 60 to 70 degree angle with the tabletop.
    4. the central ray is centered to the iliac crest.

    a. 1 and 2 only
    *b. 2 and 3 only
    c. 1, 2, and 3 only
    d. 1, 3, and 4 only

12. Hip and knee flexion for a frogleg hip radiograph:
    1. positions the greater trochanter in profile.
    2. positions the lesser trochanter in profile.
    3. rotates the greater trochanter beneath the femoral neck.
    4. determines the degree of femoral neck foreshortening.

    a. 1 and 3 only
    b. 2 and 4 only
    *c. 2 and 3 only
    d. 1 and 4 only

13. As one increases the degree of femoral abduction for a frogleg hip radiograph:
    1. the greater trochanter moves closer to the femoral head.
    2. the lesser trochanter is placed in profile.
    3. the femoral neck demonstrates increased foreshortening.
    4. the obturator foramen appears wider.

    a. 1 and 2 only
    *b. 1 and 3 only
    c. 2 and 3 only
    d. 1, 3, and 4 only

14. For the axiolateral projection of the hip:
    1. the unaffected hip should be in maximum flexion and abduction.
    2. the central ray should be positioned parallel with the femoral neck.
    3. tight collimation is needed to reduce scattered radiation.
    4. the affected leg should be internally rotated.

    a. 1 and 4 only
    b. 2 and 3 only
    *c. 1, 3, and 4 only
    d. 1, 2, 3, and 4

15. An axiolateral hip radiograph taken with the patient's affected leg in external rotation demonstrates:
    1. the greater trochanter in profile anteriorly.
    2. the greater trochanter at a transverse level halfway between the lesser trochanter and the femoral head.
    3. the greater trochanter in profile posteriorly.
    4. soft tissue from the unaffected leg superimposing the affected leg's acetabulum and femoral head.

    a. 1 only
    b. 2 and 3 only
    *c. 3 only
    d. 3 and 4 only

16. What three anatomical landmarks are used to localize the neck of the femur?
    *a. Symphysis pubis, greater trochanter, and ASIS
    b. Symphysis pubis, lesser trochanter, and ASIS
    c. Symphysis pubis, greater trochanter, and iliac crest.
    d. Symphysis pubis, greater trochanter, and ischial tuberosity.

17. The greater trochanter lies at approximately the same level as the:
    a. obturator foramen.
    b. ischial tuberosity.
    c. lesser trochanter.
    *d. symphysis pubis.

18. Internally rotating the affected leg for an axiolateral projection of the hip:
    1. positions the greater trochanter behind the femoral neck and shaft.
    2. positions the lesser trochanter in profile.
    3. positions the greater trochanter in profile.
    4. reduces the posterior decline of the femoral neck.

    a. 1 and 2 only
    b. 3 and 4 only
    c. 1, 3, and 4 only
    *d. 1, 2, and 4 only

Chapter 7

Radiographic Critique of the Cervical and Thoracic Vertebrae

I. Lecture on radiographic projections/positions of the cervical and thoracic vertebrae.

II. Lecture on radiographic evaluation for each projection/position of the cervical and thoracic vertebrae.

III. Classroom activities.

   A. Workbook study questions.

   B. Workbook learning activities numbers 2, 3, and 4.

   Suggestion for learning activity number 2: I provide several connected and unconnected cervical and thoracic bones, and accurately and poorly positioned cervical and thoracic radiographs. The students then work together, identifying the anatomical structures on the bones and on the radiographs.

   Suggestion for learning activity number 3: See chapter 2, suggestion for learning activity number 4, page 7. Fill the packets with poorly positioned cervical and thoracic radiographs.

   Suggestion for learning activity number 4: I ask the students to each bring radiographs from the cervical and thoracic vertebrae repeat bin to class. During class they label all the anatomical structures found on the radiographs.

IV. Clinical laboratory activities.

   A. Workbook learning activity number 1.

   Suggestion for learning activity number 1: I assign each student a cervical and thoracic projection/position to demonstrate to the class. See chapter 2, suggestions for learning activity number 1, page 7 for further description.

V. Workbook self test questions.

Suggestion: I ask the students not to look at or do the self test prior to completing the study question and learning activities. After the study questions and learning activities are completed the self test is taken in the classroom setting as if it were a quiz.

Used in this manner the self test is a tool that pinpoints areas of weakness before the final exam.

VI. Final exam.

The following questions pertain to both radiographic positioning and critique, and can be incorporated in your present positioning and critiquing exams. They are based on information found in the Radiographic Critique book. Since some facilities do not follow all the protocols found in the Radiographic Critique book, evaluate each question for information that does not match your facility's routines.

1. For an AP cervical radiograph:
   1. the mandibular angles, the mastoid tips, and the shoulders are positioned at equal distances to the film.
   2. the central ray is angled 15 to 20 degrees cephalically.
   3. the OML is aligned perpendicular to the film.
   4. the long axis of the cervical vertebra is aligned with the short axis of the collimated field.

   *a. 1 and 2 only
   b. 1, 2, and 3 only
   c. 2 and 3 only
   d. 3 and 4 only

2. An accurately positioned AP cervical radiograph demonstrates:
   1. each vertebra's spinous processes within the inferior adjoining vertebral body.
   2. open intervertebral disk spaces.
   3. the spinous processes aligned with the cervical bodies.
   4. the second cervical vertebra in its entirety.

   a. 1 and 2 only
   *b. 2 and 3 only
   c. 1, 2, and 3 only
   d. 1, 2, 3, and 4

3. An AP cervical radiograph taken with the patient rotated toward the right side demonstrates:
   1. the spinous processes positioned closer to the left side of the vertebral bodies.
   2. closed intervertebral joint spaces.
   3. elongation of the uncinate processes.
   4. the left SC superimposing the vertebral column.

   a. 1 only
   b. 2 and 3 only
   c. 3 and 4 only
   *d. 1 and 4 only

4. A poorly positioned AP cervical radiograph demonstrates obscured intervertebral disk spaces and each vertebra's spinous process within its vertebral body. How was the positioning setup mispositioned to obtain such a radiograph?
   a. The patient was rotated toward the right side.
   *b. The central ray was angled too caudally.
   c. The patient's head was tilted.
   d. The central ray was angled too cephalically.

5. An AP cervical radiograph taken using too much cephalic central ray angulation demonstrates:
   1. elongated uncinate processes.
   2. obscured intervertebral disk spaces.
   3. each vertebra's spinous process within the inferior adjoining vertebral body.
   4. undistorted vertebral bodies.

   a. 1 and 2 only
   b. 3 and 4 only
   *c. 1, 2, and 3 only
   d. 1, 2, 3, and 4

6. An AP cervical radiograph demonstrates the third cervical vertebra superimposed by the lower jaw. How should the positioning setup be adjusted to obtain an optimal radiograph?
   1. Decrease the degree of central ray angulation.
   2. Rotate the patient toward the left side.
   3. Elevate the chin.
   4. Align the acanthiomeatal line perpendicular to the film.

   a. 1 and 2 only
   b. 3 only
   *c. 3 and 4 only
   d. 1, 3, and 4 only

7. For an AP atlas and axis radiograph:
   1. the mandibular angles and the shoulders are positioned at equal distance to the film.
   2. the acanthiomeatal line is aligned parallel with the film.
   3. a 5-degree cephalad angulation is used.
   4. an imaginary line connecting the upper occlusal plane and posterior occiput's inferior edge is aligned perpendicular to the film.

   a. 1 and 4 only
   b. 1 and 3 only
   c. 2 and 3 only
   *d. 1, 3, and 4 only

8. An accurately positioned AP atlas and axis radiograph demonstrates:
   1. the axis's spinous process aligned with its body midline.
   2. an open atlantoaxial joint.
   3. the upper incisors and posterior occiput superior to the dens.
   4. the first through fourth cervical vertebrae.

   a. 1 and 3 only
   b. 2 only
   *c. 1, 2, and 3 only
   d. 1, 2, 3, and 4

9. An AP atlas and axis radiograph taken with the patient's face rotated toward the left side demonstrates:
   1. a narrower distance from the atlas's lateral mass to the dens on the right side than on the left side.
   2. the upper incisors obscuring the dens and atlantoaxial joint.
   3. the patient's jaw shifted toward the left side.
   4. the atlas's spinous process shifted toward the left side.

   *a. 1 and 3 only
   b. 2 only
   c. 3 and 4 only
   d. 1, 3, and 4 only

10. A poorly positioned AP atlas and axis radiograph demonstrates the upper incisors about 1 inch (2.5 cm) inferior to the posterior occiput, obscuring the dens and atlantoaxial joint. How was the positioning setup mispositioned to obtain such a radiograph?
    1. The patient's face was rotated toward the left side.
    2. The acanthiomeatal line was perpendicular to the film and the central ray was perpendicular.
    3. The patient's chin was tucked more than needed.
    4. An imaginary line connecting the occlusal plane and posterior occiput's inferior edge was not aligned perpendicular to the table.

    a. 1 only
    b. 2 only
    c. 2 and 3 only
    *d. 2, 3, and 4 only

11. A poorly positioned AP atlas and axis radiograph demonstrates the dens superimposing the posterior occiput. The upper incisors are demonstrated about 3 inches (7.5 cm) superior to the posterior occiput's inferior edge. How could the positioning setup be adjusted to obtain an optimal radiograph?
    1. Adjust the central ray that was used 15 degrees caudally.
    2. Tuck the patient's chin 1.5 inches (3.75 cm).
    3. Align the acanthiomeatal line perpendicular to the tabletop.
    4. Move the central ray and film 3 inches (7.5 cm) inferiorly.

    a. 1 and 2 only
    b. 2 and 3 only
    *c. 1, 2, and 3 only
    d. 3 and 4 only

12. The upper incisors are superimposing the dens and the posterior occiput's inferior edge is demonstrated about 1 inch (2.5 cm) superior to the dens on an AP atlas and axis radiograph. How could the positioning setup be adjusted to obtain an optimal radiograph?
    1. Align the acanthiomeatal line perpendicular to the film.
    2. Elevate the patient's chin 0.5 inch (2 cm).
    3. Adjust the central ray 5 degrees caudally.
    4. Move the central ray and film 1 inch (2.5 cm) superiorly.

    a. 1 and 2 only
    b. 2 and 3 only
    *c. 1, 2, and 3 only
    d. 4 only

13. For a lateral cervical radiograph:
    1. the midcoronal plane is positioned parallel with the film.
    2. the interpupillary line is aligned perpendicular to the film.
    3. the long axis of the cervical vertebral column is aligned with the short axis of the collimated field.
    4. a 72 inch (183 cm) SID is used.

    a. 1 and 3 only
    *b. 2 and 4 only
    c. 1, 2, and 3 only
    d. 2 and 3 only

14. An accurately positioned lateral cervical radiograph demonstrates:
    1. C-1 and C-2 without cranial or mandibular superimposition.
    2. open intervertebral disk spaces.
    3. superimposed right and left articular pillars and zygapophyseal joints.
    4. the spinous processes in profile

    a. 1 and 4 only
    b. 2 and 3 only
    c. 1, 2, and 4 only
    *d. 1, 2, 3, and 4

15. The prevertebral fat stripe:
    1. is located anterior to the cervical vertebrae.
    2. is visualized on cervical radiographs with excessive radiographic density.
    3. is used to detect fractures, masses, and inflammation within and around the cervical vertebrae.
    4. is demonstrated on a lateral cervical radiograph.

    a. 1 and 4 only
    b. 2 and 3 only
    *c. 1, 3, and 4
    d. 1, 2, 3, and 4

16. For a lateral cervical radiograph taken in maximum flexion:
    1. the patient's chin is tucked into the chest as far as possible.
    2. the intervertebral disk spaces between the cervical bodies are narrowed.
    3. the patient's chin is extended up and backwards as far as possible.
    4. the intervertebral disk spaces between the cervical bodies are widened.

    *a. 1 and 2 only
    b. 1 and 4 only
    c. 2 and 3 only
    d. 3 and 4 only

17. A poorly positioned lateral cervical radiograph demonstrates the articular pillars and zygapophyseal joints of one side of the patient situated anterior to the opposite side. How was the patient mispositioned to obtain such a radiograph?
    1. The patient was rotated.
    2. The midcoronal plane was not positioned perpendicular to the film.
    3. The head was tilted toward the film.
    4. The central ray was angled cephalically.

    a. 1 only
    *b. 1 and 2 only
    c. 2 and 3 only
    d. 4 only

18. A lateral cervical radiograph taken with the patient's head tilted toward the film demonstrates:
    1. the inferior cortices of the mandibular without superimposition.
    2. the articular pillars and zygapophyseal joints without superimposition.
    3. the vertebral foramen of C-1.
    4. superimposed inferior cranial cortices.

    a. 1 only
    *b. 1 and 3 only
    c. 2 and 4 only
    d. 3 and 4 only

19. The vertebral body of C-7 is not demonstrated on a lateral cervical radiograph. To demonstrate this cervical vertebra:
    1. take the radiograph on expiration.
    2. have the patient hold weights on each arm to depress shoulders.
    3. take a swimmers lateral radiograph.
    4. angle the central ray 10 degrees cephalically.

    a. 1 and 2 only
    b. 2 and 3 only
    *c. 1, 2, and 3 only
    d. 3 and 4 only

20. For an anterior oblique cervical radiograph:
    1. the midcoronal plane is aligned at a 45-degree angle with the film.
    2. the central ray is angled 15 degrees caudally.
    3. the head's midsagittal place is aligned perpendicular to the film.
    4. a 40 inch (102 cm) SID is used.

    *a. 1 and 2 only
    b. 1 and 3 only
    c. 2 and 3 only
    d. 1, 2, and 4 only

21. An accurately positioned oblique cervical radiograph demonstrates:
    1. the zygapophyseal joints.
    2. the intervertebral foramina.
    3. open intervertebral disk spaces.
    4. the inferior cortical outlines of the mandible without superimposition.

    a. 1 only
    b. 2 and 4 only
    c. 2 and 3 only
    *d. 2, 3, and 4 only

22. Which of the following positions demonstrate the right intervertebral foramina?
    1. lateral
    2. LPO
    3. LAO
    4. RAO

    a. 1 only
    b. 2 only
    c. 2 and 3 only
    *d. 2 and 4 only

23. An LAO cervical radiograph taken with the patient over rotated demonstrates:
    1. the right pedicles in the midlines of the vertebral bodies.
    2. the right pedicles in profile.
    3. the left zygapophyseal joints are visualized.
    4. the vertebral column superimposes the right SC joint and medial clavicle.

    a. 1 and 4 only
    *b. 1 and 3 only
    c. 1, 2, and 3 only
    d. 3 and 4 only

24. An oblique cervical radiograph demonstrates closed intervertebral disk spaces and distorted vertebral bodies when:
   *a. the central ray angulation is inaccurate.
   b. the patient is over rotated.
   c. the patient's head is not in a lateral position.
   d. the patient is under rotated.

25. For a swimmers cervical radiograph:
   1. the patient is placed in a lateral position.
   2. the arm placed adjacent to the tabletop is elevated.
   3. the arm placed farthest from the tabletop is positioned at a 90-degree angle with the body.
   4. the midsagittal plane is aligned parallel with the film.

   a. 1 and 4 only
   b. 2 and 3 only
   *c. 1, 2, and 4 only
   d. 1, 2, 3, and 4

26. A poorly positioned swimmers cervical radiograph demonstrates the humerus with the greatest degree of magnification rotated posteriorly. How was the patient mispositioned to obtain such a radiograph?
   *a. The arm situated farthest from the film was rotated posteriorly.
   b. The midcoronal plane was not aligned parallel with the film.
   c. The arm situated closest to the film was rotated posteriorly.
   d. The arm situated closest to the film was not elevated.

27. An accurately positioned swimmers cervical radiograph demonstrates:
   1. distorted vertebral bodies.
   2. superimposed right and left articular pillars.
   3. open intervertebral disk spaces.
   4. the fifth through seventh cervical and the first through third thoracic vertebrae.

   a. 1 and 4 only
   b. 2 and 3 only
   c. 3 and 4 only
   *d. 2, 3, and 4 only

28. A swimmers cervical radiograph demonstrates closed intervertebral disk spaces and distorted vertebral bodies. How was the patient mispositioned to obtain such a radiograph?
   a. The midcoronal plane was not aligned parallel with the film.
   b. The arm situated farthest away from the film was not depressed.
   c. The patient was rotated anteriorly.
   *d. The midsagittal plane was not aligned parallel with the film.

29. For an AP thoracic radiograph:
   1. the shoulders and the ASISs are positioned at equal distances from the tabletop.
   2. the hips and knees are flexed until the lower back is pressed against the tabletop.
   3. the central ray is centered to the fifth thoracic vertebra.
   4. the transversely collimated field is open to an 8 inch (20 cm) field size.

   a. 1 and 2 only
   b. 3 and 4 only
   *c. 1, 2, and 4 only
   d. 1, 2, 3, and 4

30. An accurately positioned AP thoracic radiograph demonstrates:
    1. the vertebral bodies are distorted.
    2. the long axis of the thoracic vertebrae is aligned with the long axis of the collimated field.
    3. the spinous processes are aligned with the midline of the vertebral bodies.
    4. the intervertebral disk spaces are open.

    a. 1 and 4 only
    b. 2 and 3 only
    *c. 2, 3, and 4 only
    d. 1, 2, 3, and 4

31. To obtain uniform density throughout the entire thoracic vertebrae on an AP thoracic radiograph:
    1. position a wedge type compensating filter over the upper thoracic vertebrae.
    2. use a high ratio grid.
    3. position the patient's feet toward the cathode end of the tube.
    4. use a low kVp technique.

    a. 1 only
    b. 2 and 4 only
    *c. 1 and 3 only
    d. 3 only

32. A poorly positioned AP thoracic radiograph demonstrates closed eighth through twelfth intervertebral disk spaces. How was the patient mispositioned to obtain such a radiograph?
    a. the patient was rotated toward the left side.
    *b. the patient's knees and hips were extended.
    c. the long axis of the vertebral column was laterally tilted.
    d. the patient's head was on a thick pillow.

33. An AP thoracic radiograph taken with the patient rotated toward the left side demonstrates:
    1. the spinous processes positioned closer to the right side.
    2. a greater distance from the right pedicle to the spinous process than from the left pedicle to the spinous process.
    3. the right SC joint superimposing the vertebral column.
    4. closed intervertebral disk spaces.

    *a. 1 and 3 only
    b. 2 and 3 only
    c. 1, 2, and 3 only
    d. 4 only

34. For a lateral thoracic radiograph:
    1. the shoulders are positioned at equal distances to the tabletop.
    2. the arms are abducted to a 90-degree angle.
    3. breathing technique is used.
    4. a radiolucent sponge is positioned between the patient's lateral body surface and the tabletop at a level just inferior to the iliac crest.

    a. 1, and 4 only
    *b. 2 and 3 only
    c. 1, 2, and 3 only
    d. 2, 3, and 4

35. An accurately positioned lateral thoracic radiograph demonstrates:
    1. the intervertebral foramina.
    2. about 0.5 inch (1 cm) of space between the posterior ribs.
    3. open intervertebral disk spaces.
    4. the pedicles in profile.

    a. 1 and 3 only
    b. 2 and 4 only
    c. 1, 2, and 3 only
    *d. 1, 2, 3, and 4

36. A poorly positioned lateral thoracic radiograph demonstrates more than 0.5 inch (1 cm) of space visualized between the posterior ribs. How should the patient be repositioned to obtain an optimal radiograph?
    1. Rotate the patient's elevated side anteriorly.
    2. Position the midsagittal plane perpendicular to the tabletop.
    3. Center the central ray and film superiorly.
    4. Align shoulders, posterior ribs, and posterior pelvic wings perpendicular to the tabletop.

    a. 1 only
    *b. 1 and 4 only
    c. 2 only
    d. 2 and 3 only

37. A poorly positioned lateral thoracic radiograph demonstrates closed eighth through twelfth intervertebral disk spaces. How was the patient mispositioned to obtain such a radiograph?
    1. The central ray was not aligned perpendicular to the thoracic vertebrae.
    2. The vertebral column was not aligned parallel with the tabletop.
    3. A radiolucent sponge was not accurately positioned between the lateral surface of the patient and the tabletop.
    4. The hips and knees were extended.

    a. 1 and 2 only
    b. 2 and 3 only
    *c. 1, 2, and 3 only
    d. 4 only

Chapter 8

Radiographic Critique of the Lumbar Vertebrae, Sacrum, and Coccyx

I. Lecture on radiographic projections/positions of the lumbar vertebrae, sacrum, and coccyx.
II. Lecture on radiographic evaluation for each projection/position of the lumbar vertebrae, sacrum, and coccyx.
III. Classroom activities.

   A. Workbook study questions.

   B. Workbook learning activities numbers 2, 3, and 4.

   Suggestion for learning activity number 2: I provide several connected and unconnected lumbar, sacral, and coccygeal bones, and accurately and poorly positioned lumbar, sacral, and coccygeal radiographs. The students then work together, identifying the anatomical structures on the bones and on the radiographs.

Suggestion for learning activity number 3: See chapter 2, suggestion for learning activity number 4, page 7. Fill the packets with poorly positioned lumbar, sacral, and coccygeal radiographs.

Suggestion for learning activity number 4: I ask the students to each bring radiographs from the lumbar, sacral, and coccygeal repeat bin to class. During class they label all the anatomical structures found on the radiographs.

IV. Clinical laboratory activities.
   A. Workbook learning activity number 1.

   Suggestion for learning activity number 1: I assign each student a lumbar, sacral, or coccygeal projection/position to demonstrate to the class. See chapter 2, suggestions for learning activity number 1, page 7 for further description.

V. Workbook self test questions.

   Suggestion: I ask the students not to look at or do the self test prior to completing the study question and learning activities. After the study questions and learning activities are completed the self test is taken in the classroom setting as if it were a quiz.
   Used in this manner the self test is a tool that pinpoints areas of weakness before the final exam.

VI. Final exam.
   The following questions pertain to both radiographic positioning and critique, and can be incorporated in your present positioning and critiquing exams. They are based on information found in the Radiographic Critique book. Since some facilities do not follow all the protocols found in the Radiographic Critique book, evaluate each question for information that does not match your facility's routines.

1. For an AP lumbar radiograph:
   1. the ASIS's are positioned at equal distances to the tabletop.
   2. the patient's legs are extended.
   3. the long axis of the vertebral column is aligned with the short axis of the collimated field.
   4. the central ray is centered to the iliac crest when a 14 x 17 inch (43 x 35 cm) film is used.

   *a. 1 and 4 only
   b. 2 only
   c. 1, 2, and 4 only
   d. 1, 2, 3, and 4

2. An accurately positioned AP lumbar radiograph demonstrates:
   1. open intervertebral disk spaces.
   2. the vertebral bodies without distortion.
   3. the spinous processes aligned with the midline of the vertebral bodies.
   4. the sacrum and coccyx are aligned with the symphysis pubis.

   a. 1 and 2 only
   b. 3 and 4 only
   c. 1, 3, and 4 only
   *d. 1, 2, 3, and 4

3. An AP lumbar radiograph taken with the patient rotated toward the right side demonstrates:
   1. the spinous processes positioned closer to the left pedicles.
   2. the sacrum and coccyx rotated toward the right side.
   3. closed intervertebral disk spaces.
   4. distorted vertebral bodies.

   *a. 1 only
   b. 1 and 2 only
   c. 1 and 3 only
   d. 3 and 4 only

4. A poorly positioned AP lumbar radiograph demonstrating closed intervertebral disk spaces:
   1. also demonstrates distorted vertebral bodies.
   2. was taken with the patient rotated.
   3. was taken with the patient's legs extended.
   4. also demonstrates the sacrum and coccyx rotated toward the left side.

   a. 1 and 2 only
   b. 2 and 3 only
   *c. 1 and 3 only
   d. 2 and 4 only

5. For a posterior oblique lumbar radiograph:
   1. the patient's thorax is rotated until the midcoronal plane is at a 40 to 45 degree angle with film.
   2. the patient's pelvis is rotated until the midcoronal plane is at a 55 to 60 degree angle with film.
   3. the central ray is centered 2 inches (5 cm) medial to the elevated ASIS at a level 1 to 1.5 inches (2.5 to 4 cm) superior to the iliac crest.
   4. the long axis of the vertebral column is aligned with the short axis of the collimated field.

   a. 1 and 3 only
   b. 2 and 4 only
   *c. 1, 2, and 3 only
   d. 1, 2, 3, and 4

6. An accurately positioned posterior oblique lumbar radiograph demonstrates:
   1. the superior and inferior articular processes in profile.
   2. "Scotty dogs" that are stacked on top of one another.
   3. the obturator foramina.
   4. the pedicles situated closest to the film in the center of the vertebral bodies.

   a. 1 and 2 only
   b. 3 only
   *c. 1, 2, and 4 only
   d. 1, 2, 3, and 4

7. A posterior oblique lumbar radiograph taken with the patient under rotated demonstrates:
   1. obscured zygapophyseal joints.
   2. the pedicles closer to the lateral surface of the vertebral bodies than the midline.
   3. the intervertebral foramina.
   4. the inferior and superior articular processes in profile.

   *a. 1 and 2 only
   b. 2 and 3 only
   c. 3 only
   d. 1, 2, and 4 only

8. For a lateral lumbar radiograph:
   1. the vertebral column is aligned parallel with the tabletop.
   2. align the shoulders, the posterior ribs, and the posterior pelvic wings perpendicular to the tabletop.
   3. on a scoliotic patient, the patient is positioned on the tabletop so the central ray is directed into the spinal curve.
   4. the patient's knees are flexed and a pillow or radiolucent sponge is placed between them.

   a. 1 and 2 only
   b. 3 and 4 only
   c. 1, 2, and 4 only
   *d. 1, 2, 3, and 4

9. An accurately positioned lateral lumbar radiograph demonstrates:
   1. open intervertebral disk spaces.
   2. the intervertebral foramina.
   3. distorted vertebral bodies.
   4. the zygapophyseal joints.

   *a. 1 and 2 only
   b. 3 and 4 only
   c. 1, 2, and 3 only
   d. 1 and 4 only

10. For a lateral lumbar radiograph taken in maximum extension:
    1. the patient arches the back by extending the shoulders and legs as far posteriorly as possible.
    2. the lumbar vertebral column demonstrates an increase in lordotic curvature.
    3. the lumbar vertebral column demonstrates a decrease in lordotic curvature.
    4. the patient rolls into a tight ball.

    *a. 1 and 2 only
    b. 1 and 3 only
    c. 2 and 4 only
    d. 3 and 4 only

11. A poorly positioned left lateral lumbar radiograph demonstrates rotation. The posterior ribs that are most magnified and projected inferiorly are rotated anteriorly. How should the patient be repositioned to obtain an optimal radiograph?
    *a. Rotate the patient's right side posteriorly.
    b. Angle the central ray caudally.
    c. Rotate the patient's left side posteriorly
    d. Position the vertebral column parallel with the tabletop.

12. A lateral lumbar radiograph demonstrates closed L4-5 and L5-S1 intervertebral disk spaces. How should the positioning setup be adjusted to obtain an optimal radiograph?
    1. Angle the central ray 5 to 8 degrees caudally.
    2. Rotate the patient into a true lateral position.
    3. Align the central ray perpendicular to the vertebral column.
    4. Align the vertebral column parallel with the tabletop.

    a. 1 and 4 only
    b. 2 only
    c. 3 and 4 only
    *d 1, 3, and 4 only

13. For a lateral L5-S1 spot radiograph:
    1. the vertebral column is aligned parallel with the tabletop.
    2. a high kVp is used to penetrate the hips and pelvis.
    3. the shoulders and the ASISs are positioned at equal distances to the tabletop.
    4. the knees are flexed.

    a. 1 and 2 only
    b. 3 and 4 only
    *c. 1, 2, and 4 only
    d. 1, 2, 3, and 4

14. An accurately positioned lateral L5-S1 spot radiograph demonstrates:
    1. the L5-S1 intervertebral disk space in the center of the collimated field.
    2. an open L5-S1 intervertebral disk space.
    3. obscured intervertebral foramina.
    4. nearly superimposition of the greater sciatic notches.

    a. 1 and 2 only
    b. 2 and 3 only
    c. 1 and 4 only
    *d. 1, 2, and 4 only

15. For an AP sacral radiograph:
    1. the patient should empty the bladder and colon prior to the procedure.
    2. the central ray is angled 15 degrees cephalically.
    3. the shoulders and the ASISs are positioned at equal distance to the tabletop.
    4. the hips and knees are flexed.

    a. 1 only
    b. 2 and 3 only
    *c. 1, 2, and 3 only
    d. 1, 2, 3, and 4

16. An accurately positioned AP sacral radiograph demonstrates:
    1. the long axis of the median crest aligned with the long axis of the collimated field.
    2. foreshortening of the first through fifth sacral segments.
    3. the median sacral crest is positioned closer to the right side.
    4. the ischial spines are equally demonstrated and are aligned with the pelvic brim.

    a. 1 and 2 only
    *b. 1 and 4 only
    c. 2 and 3 only
    d. 1, 2, and 4 only

17. A poorly positioned AP sacral radiograph demonstrates the symphysis pubis rotated toward the patient's right side. How was the positioning setup mispositioned to obtain such a radiograph?
    a. The central ray was angled too cephalically.
    *b. The patient was in an RPO position.
    c. The patient's legs were extended.
    d. The patient was in an LPO position.

18. An AP sacral radiograph taken with a perpendicular central ray demonstrates:
    1. an elongated sacrum.
    2. the sacrum and symphysis pubis without alignment.
    3. a foreshortened sacrum.
    4. the symphysis pubis superimposing the lower sacral segments.

    a. 1 and 4 only
    b. 2 only
    c. 3 only
    *d. 3 and 4 only

19. For a lateral sacral radiograph:
    1. the patient's legs are flexed.
    2. the midcoronal plane is aligned parallel with the tabletop.
    3. the posterior ribs and the posterior pelvic wings are aligned perpendicular to the tabletop.
    4. the longitudinal axis of the sacrum is aligned with the short axis of the collimated field.

    *a. 1 and 3 only
    b. 2 and 3 only
    c. 1, 2, and 3 only
    d. 2 and 4 only

20. An accurately positioned lateral sacral radiograph demonstrates:
    1. the long axis of the sacrum aligned with the long axis of the collimated field.
    2. near superimposition of the greater sciatic notches.
    3. an open L5-S1 intervertebral disk space.
    4. the median sacral crest is in profile.

    a. 1 and 2 only
    b. 1, 2, and 3 only
    c. 3 and 4 only
    *d. 1, 2, 3, and 4

21. A left lateral sacral radiograph demonstrates the greater sciatic notches without superimposition and the superiorly situated femoral head visualized anteriorly. How should the positioning setup be adjusted to obtain an optimal radiograph?
    *a. Rotate the right side of the pelvis anteriorly.
    b. Position a radiolucent sponge between the lateral body surface and the tabletop to eliminate lumbar sagging.
    c. Angle the central ray caudally.
    d. Rotate the left side of the pelvis anteriorly.

22. A poorly positioned lateral sacral radiograph demonstrates a closed L5-S1 intervertebral disk space and foreshortening of the sacrum. How could the positioning setup have been mispositioned to obtain such a radiograph?
    1. The patient was rotated.
    2. The central ray was angled too cephalically.
    3. The vertebral column was not aligned parallel with the tabletop.
    4. The patient's legs were extended.

    a. 1 only
    *b. 2 and 3 only
    c. 3 only
    d. 1 and 4 only

23. For an AP coccygeal radiograph:
    1. the central ray is angled 10 degrees cephalically.
    2. the shoulders and the ASIS's are positioned perpendicular to the tabletop.
    3. it is suggested the patient empties the bladder and colon prior to the exam.
    4. the central ray is centered to the midsagittal plane at a level 2 inches (5 cm) superior to the symphysis pubis.

    a. 1 and 3 only
    b. 2 and 4 only
    *c. 3 and 4 only
    d. 1, 2, 3, and 4 only

24. An accurately positioned AP coccygeal radiograph demonstrates:
    1. the coccyx aligned with the symphysis pubis.
    2. the longitudinal axis of the coccyx aligned with the longitudinal axis of the film.
    3. the first through third coccygeal vertebrae.
    4. the coccyx without foreshortening.

    a. 1 and 3 only
    b. 2 and 3 only
    c. 1, 2, and 4 only
    *d. 1, 2, 3, and 4

25. An AP coccygeal radiograph demonstrates the symphysis pubis superimposing second and third coccygeal vertebrae. How was the positioning setup mispositioned to obtain such a radiograph?
    a. The patient was rotated.
    b. The central ray was angled too caudally.
    *c. The central ray was angled too cephalically.
    d. The patient's legs were extended.

26. For a lateral coccygeal radiograph:
    1. the longitudinally and transversely collimated fields can safely be closed to a 4 inch (10 cm) field size.
    2. the posterior ribs and the posterior pelvic wings are positioned perpendicular to the tabletop.
    3. the patient's legs are extended.
    4. a small focal spot improves recorded detail sharpness.

    a. 1 and 4 only
    b. 2 and 3 only
    *c. 1, 2, and 4 only
    d. 1, 2, 3, and 4

27. An accurately positioned left lateral coccygeal radiograph demonstrates:
    1. the median sacral crest in profile.
    2. the first coccygeal vertebra in the center of the collimated field.
    3. nearly superimposed greater sciatic notches.
    4. a left marker.

    a. 1 and 4 only
    b. 2 and 3 only
    c. 2, 3, and 4 only
    *d. 1, 2, 3, and 4

Chapter 9

Radiographic Critique of the Sternum and Ribs

I.   Lecture on radiographic projections/positions of the sternum and ribs.

II. Lecture on radiographic evaluation for each projection/position of the sternum and ribs.

III. Classroom activities.

   A. Workbook study questions.

   B. Workbook learning activities numbers 2, 3, and 4.

   Suggestion for learning activity number 2: I provide a skeletal thorax, and accurately and poorly positioned sternal and rib radiographs. The students then work together, identifying the anatomical structures on the bones and on the radiographs.

Suggestion for learning activity number 3: See chapter 2, suggestion for learning activity number 4, page 7. Fill the packets with poorly positioned sternal and rib radiographs.

Suggestion for learning activity number 4: I ask the students to each bring radiographs from the sternal and rib repeat bin to class. During class they label all the anatomical structures found on the radiographs.

IV. Clinical laboratory activities.

   A. Workbook learning activity number 1.

   Suggestion for learning activity number 1: I assign each student a sternal and rib projection/position to demonstrate to the class. See chapter 2, suggestions for learning activity number 1, page 7 for further description.

V. Workbook self test questions.

   Suggestion: I ask the students not to look at or do the self test prior to completing the study question and learning activities. After the study questions and learning activities are completed the self test is taken in the classroom setting as if it were a quiz.
   Used in this manner the self test is a tool that pinpoints areas of weakness before the final exam.

VI. Final exam.
   The following questions pertain to both radiographic positioning and critique, and can be incorporated in your present positioning and critiquing exams. They are based on information found in the Radiographic Critique book. Since some facilities do not follow all the protocols found in the Radiographic Critique book, evaluate each question for information that does not match your facility's routines.

1. For an RAO sternal radiograph:
   1. a 30 inch (76 cm) SID is used.
   2. the patient's midcoronal plane is angled 15 to 20 degrees with the film.
   3. a short exposure time is used.
   4. the radiograph is taken on expiration.

   *a. 1 and 2 only
   b. 1 and 3 only
   c. 2 and 3 only
   d. 1, 2, and 4 only

2. On an accurately positioned RAO sternal radiograph:
   1. the manubrium is demonstrated to the left of the heart shadow.
   2. the posterior ribs are magnified.
   3. the sternum is visualized about 3 inches (7.5 cm) to the left of the thoracic vertebral column.
   4. the lung markings are blurred.

   a. 1 and 3 only
   b. 2 and 3 only
   c. 1, 2, and 4 only
   *d. 2, 3, and 4 only

3. A poorly positioned RAO sternal radiograph demonstrates the right SC joint and manubrium superimposed by the thoracic vertebrae. How should the patient be repositioned to obtain an optimal radiograph?
   a. Have patient take a deeper inspiration before taking radiograph.
   b. Center the central ray and film closer to the vertebral column.
   *c. Increase the degree of patient obliquity.
   d. Decrease the degree of patient obliquity.

4. On an RAO sternal radiograph the posterior ribs are blurred and magnified because:
   1. a 30 inch (76 cm) SID is used.
   2. a long exposure time is used.
   3. the radiograph is taken on expiration.
   4. a detail screen is used.

   *a. 1 and 2 only
   b. 1, 2, and 3 only
   c. 1 only
   d. 2 and 4 only

5. A lateral sternal radiograph:
   1. is taken with the shoulders positioned at equal distances to the film.
   2. demonstrates the sternum without humeral soft tissue superimposition.
   3. is taken upon deep inspiration.
   4. is taken with the central ray centered halfway between the sternal angle and the xiphoid process.

   a. 1, 2, and 4 only
   *b. 2 and 3 only
   c. 3 only
   d. 2 and 4 only

6. A poorly positioned left lateral sternal radiograph demonstrates the superior heart shadow extending beyond the sternum into the anteriorly located lung. How should the patient be repositioned to obtain an optimal radiograph?
   1. Rotate the left thorax posteriorly.
   2. Position the arms behind the patient's back.
   3. Take the exposure on deep inspiration.
   4. Rotate the right thorax anteriorly.

   a. 1 only
   *b. 1 and 4 only
   c. 2 and 3 only
   d. 3 and 4 only
   e. 4 only

7. An AP/PA rib radiograph on a patient with lower anterior rib pain is taken:
   1. with the patient in an AP projection.
   2. on expiration.
   3. using 65 to 70 kVp.
   4. with the shoulders at equal distances to the tabletop.

   a. 1 and 3 only
   *b. 2 and 4 only
   c. 1 and 4 only
   d. 1, 2, 3, and 4

8. An accurately positioned above diaphragm AP/PA rib radiograph demonstrates:
   1. the scapulae outside the lung field.
   2. the seventh posterior rib at the center of the collimated field.
   3. the 9th through 12th posterior ribs below the diaphragm.
   4. more distance from the right SC joint to the vertebral column than from the left SC joint to the vertebral column.

   *a. 1 and 2 only
   b. 3 and 4 only
   c. 2 and 3 only
   d. 1 and 4 only

9. For a posterior oblique rib radiograph taken to evaluate upper posterior rib pain:
   1. a 65 to 70 kVp technique is used.
   2. the central ray is centered at a level halfway between the jugular notch and xiphoid.
   3. the patient is rotated 45 degrees away from the affected side.
   4. the radiograph is taken after deep inspiration.

   a. 1 and 2 only
   b. 1 and 4 only
   *c. 1, 2, and 4 only
   d. 1, 2, 3, and 4

10. An accurately positioned below diaphragm RPO rib radiograph demonstrates:
    1. the 9th through 12th ribs below the diaphragm.
    2. layered out axillary ribs.
    3. the seventh axillary rib at the center of the collimated field.
    4. the inferior sternal body just to the right of the vertebral column.

    *a. 1 and 2 only
    b. 2 and 4 only
    c. 1 and 3 only
    d. 1, 2, and 4 only

11. An oblique rib radiograph taken with the patient rotated less than 45 degrees demonstrates:
    1. posterior ribs that are layered out.
    2. axillary ribs that are self-superimposed.
    3. anterior ribs that are demonstrated adjacent to the lateral edge of the film.
    4. the sternal body next to the vertebral column.

    a. 1 and 2 only
    b. 2 and 4 only
    c. 3 and 4 only
    *d. 1, 2, and 4 only

Chapter 10

Radiographic Critique of the Cranium

I. Lecture on radiographic projections/positions of the cranium.

II. Lecture on radiographic evaluation for each projection/position of the cranium.

III. Classroom activities.

   A. Workbook study questions.

   B. Workbook learning activities numbers 2, 3, and 4.

   Suggestion for learning activity number 2: I provide several cranial bones, and accurately and poorly positioned cranial radiographs. The students then work together, identifying the anatomical structures on the bones and on the radiographs.

   Suggestion for learning activity number 3: See chapter 2, suggestion for learning activity number 4, page 7. Fill the packets with poorly positioned cranial radiographs.

   Suggestion for learning activity number 4: I ask the students to each bring radiographs from the cranial repeat bin to class. During class they label all the anatomical structures found on the radiographs.

IV. Clinical laboratory activities.

   A. Workbook learning activity number 1.

   Suggestion for learning activity number 1: I assign each student a cranial projection/position to demonstrate to the class. See chapter 2, suggestions for learning activity number 1, page 7 for further description.

V. Workbook self test questions.

   Suggestion: I ask the students not to look at or do the self test prior to completing the study questions and learning activities. After the study questions and learning activities are completed the self test is taken in the classroom setting as if it were a quiz.
   Used in this manner the self test is a tool that pinpoints areas of weakness before the final exam.

VI. Final exam.
   The following questions pertain to both radiographic positioning and critique, and can be incorporated in your present positioning and critiquing exams. They are based on information found in the Radiographic Critique book. Since some facilities do not follow all the protocols found in the Radiographic Critique book, evaluate each question for information that does not match your facility's routines.

1. For a PA cranial radiograph:
   1. the midsagittal plane is positioned parallel with the film.
   2. the OML is aligned perpendicular to the film.
   3. the central ray is aligned perpendicular to the film.
   4. the central ray is centered to the nasion.

   a. 1 and 3 only
   *b. 2 and 3 only
   c. 2, 3, and 4 only
   d. 1 and 4 only

2. An accurately positioned PA cranial radiograph demonstrates:
   1. equal distances from the oblique orbital lines to the lateral cranial cortices on each side.
   2. the anterior clinoids and dorsum sellae superior to the ethmoid sinuses.
   3. the petrous ridges horizontally through the lower one-third of the orbits.
   4. the cristi galli and nasal septum aligned with the long axis of the film.

   a. 1 and 2 only
   b. 3 only
   *c. 1, 2, and 4 only
   d. 1, 2, 3, and 4

3. An AP cranial radiograph can be distinguished from a PA cranial radiograph because:
   1. it demonstrates less orbital magnification.
   2. it demonstrates the internal auditory canals horizontally through the orbits.
   3. it demonstrates less distance from the obliques orbital lines to the lateral cranial cortices.
   4. it demonstrates the anterior clinoids and dorsum sellae superior to the ethmoid sinuses.

   a. 1 only
   b. 1 and 3 only
   *c. 3 only
   d. 2, 3, and 4 only

4. A PA cranial radiograph taken with the patient's face rotated toward the right side demonstrates:
   *a. a greater distance from the oblique orbital line to the lateral cranial cortex on the left side than on the right side.
   b. the petrous ridges inferior to the supraorbital rims.
   c. the petrous ridges superior to the supraorbital rims.
   d. a greater distance from the oblique orbital line to the lateral cranial cortex on the right side than on the left side.

5. A poorly positioned PA cranial radiograph demonstrates the petrous ridges inferior to the supraorbital rims. How was the patient mispositioned to obtain such a radiograph?
   1. The patient's chin was not adequately tucked.
   2. The OML was not positioned perpendicular to the film.
   3. The patient's face was rotated toward the right side.
   4. The patient's head was tilted.

   *a. 1 and 2 only
   b. 2 only
   c. 2 and 3 only
   d. 4 only

6. A poorly positioned AP cranial radiograph demonstrates the petrous ridges inferior to the supraorbital rims. How could the positioning setup be adjusted to obtain an optimal radiograph?
   1. Rotate the patient's face toward the left side.
   2. Increase the degree of caudal central ray angulation.
   3. Position the OML perpendicular to the film.
   4. Tuck the patient's chin more.

   a. 1 only
   b. 2 and 3 only
   c. 3 and 4 only
   *d. 2, 3, and 4 only

7. For a PA (Caldwell) cranial radiograph:
   1. the midsagittal plane is aligned perpendicular to the film.
   2. the central ray is angled 15 degrees caudally.
   3. the OML is aligned perpendicular to the film.
   4. the central ray is centered to the nasion.

   a. 1 and 3 only
   b. 2 and 4 only
   c. 2, 3, and 4 only
   *d. 1, 2, 3, and 4

8. An accurately positioned PA (Caldwell) cranial radiograph:
   1. equal distance from the crista galli to the lateral cranial cortices on each side.
   2. the petrous ridges aligned with the supraorbital rims.
   3. the petrous pyramids superimposing the infraorbital rims.
   4. the superior orbital fissures demonstrated within the orbits.

   a. 1 and 4 only
   b. 2 only
   *c. 1, 3, and 4 only
   d. 1, 2, 3, and 4

9. A poorly positioned PA (Caldwell) cranial radiograph demonstrates the petrous ridges inferior to the infraorbital rims. How was the patient positioned to obtain such a radiograph?
   1. The patient's chin was not tucked enough.
   2. The OML was not positioned perpendicular to the film.
   3. The patient's face was rotated toward the right side.
   4. The patient's head was tilted.

   *a. 1 and 2 only
   b. 2 only
   c. 2 and 3 only
   d. 4 only

10. A poorly positioned AP (Caldwell) cranial radiograph demonstrates the petrous ridges inferior to the infraorbital rims. How could the positioning setup be adjusted to obtain an optimal radiograph?
    1. Rotate the patient's face toward the left side.
    2. Increase the degree of caudal central ray angulation.
    3. Position the OML perpendicular to the film.
    4. Elevate the patient's chin.

    a. 1 only
    *b. 2 and 3 only
    c. 3 and 4 only
    d. 2, 3, and 4 only

11. When the central ray is aligned with an AP patient's OML the tube angle reads 13 degrees caudad. What angulation should be used for this patient to obtain accurate anatomical alignment for a trauma AP (Caldwell) cranial radiograph?
    a. 15 degrees caudad
    b. 2 degrees caudad
    *c. 2 degrees cephalad
    d. perpendicular

12. For an AP Axial (Towne) radiograph:
    1. the midsagittal plane is aligned with the long axis of the collimator's longitudinal axis.
    2. the midsagittal plane is positioned parallel with the film.
    3. the central ray is angled 30 degrees caudally.
    4. the OML is aligned perpendicular to the film.

    a. 1 and 2 only
    b. 3 and 4 only
    *c. 1, 3, and 4 only
    d. 1, 2, 3, and 4

13. An accurately positioned AP Axial (Towne) radiograph demonstrates:
    1. equal distances from the posterior clinoid processes to the lateral borders of the foramen magnum on each side.
    2. the sagittal suture and nasal septum aligned with the long axis of the collimated field.
    3. the dorsum sellae within the foramen magnum.
    4. the posterior clinoids inferior to the foramen magnum.

    a. 1, 2, and 3 only
    *b. 1 and 3 only
    c. 2 and 4 only
    d. 1, 2, 3, and 4

14. When the central ray is aligned with an AP patient's OML the tube angle reads 20 degrees caudad. What angulation should be used for this patient to obtain accurate anatomical alignment for a trauma AP Axial (Towne) cranial radiograph?
    a. perpendicular
    b. 50 degrees caudad
    *c. 45 degrees caudad
    d. 10 degrees cephalad

15. Which of the following pertains to an AP Axial (Towne) cranial radiograph taken with the patient's face rotated toward the left side?
    a. The dorsum sellae is demonstrated superior to the foramen magnum.
    b. The atlas's posterior arch is demonstrated within the foramen magnum.
    *c. The distance from the dorsum sellae to the lateral foramen magnum on the patient's left side is narrower than the same distance on the right side.
    d. The distance from the dorsum sellae to the lateral foramen magnum on the patient's right side is narrower than the same distance on the left side.

16. A poorly positioned AP Axial (Towne) cranial radiograph demonstrates the dorsum sellae superior to the foramen magnum. How was the position setup mispositioned to obtain such a radiograph?
    1. The patient's face was rotated toward the left side.
    2. The chin was not adequately tucked.
    3. The OML was not aligned perpendicular to the film.
    4. The central ray was angled too caudally.

    a. 1 only
    b. 2 only
    *c. 2 and 3 only
    d. 2, 3, and 4 only

17. A poorly positioned AP Axial (Towne) cranial radiograph demonstrates a foreshortened dorsum sellae and the atlas's posterior arch within the foramen magnum. How should the positioning setup be adjusted to obtain an optimal radiograph?
    1. The patient's face was rotated toward the right side.
    2. The chin was not adequately tucked.
    3. The OML was not aligned perpendicular to the film.
    4. The central ray was angled too caudally.

    a. 1 only
    b. 4 only
    *c. 3 and 4 only
    d. 2, 3, and 4 only

18. For a lateral cranial radiograph:
    1. the midsagittal plane is positioned parallel with the film.
    2. the interpupillary line is positioned parallel with the film.
    3. the infraorbitomeatal line is positioned perpendicular to the front edge of the cassette.
    4. the central ray is centered 2 inches (5 cm) anterior to the EAM.

    *a. 1 and 3 only
    b. 1 and 4 only
    c. 2 and 3 only
    d. 1, 3, and 4

19. An accurately positioned lateral cranial radiograph demonstrates:
    1. the sellae turcica in profile.
    2. the right orbital roof slightly superior to the left orbital roof.
    3. the dorsum sellae within the foramen magnum.
    4. superimposed mandibular rami.

    *a. 1 and 4 only
    b. 3 only
    c. 1, 2 and 4 only
    d. 4 only

20. A lateral cranial radiograph demonstrates the external auditory meatus and the inferior cranial cortices without superimposition. One of each corresponding structure is visualized inferior to the other. How was the patient mispositioned to obtain such a radiograph?
    *a. The patient's head was tilted.
    b. The patient's head was rotated.
    c. The patient's chin was elevated.
    d. The central ray was centered too superiorly.

21. For a submentovertex (Basilar) cranial radiograph:
    1. the central ray is aligned perpendicular to the film.
    2. the central ray is centered to the midcoronal plane at a level 0.5 inch (1.25 cm) posterior to the mandibular angles.
    3. the IOML is parallel with the film.
    4. the midsagittal plane is perpendicular to the film.

    a. 1 and 2 only
    b. 3 and 4 only
    c. 1, 3 and 4 only
    *d. 1, 2, 3, and 4

22. A poorly positioned lateral cranial radiograph demonstrates the greater wings of the sphenoid and the anterior cranial cortices without superimposition. One of each corresponding structure is visualized posterior to the other. How was the patient mispositioned to obtain such a radiograph?
    a. The patient's head was tilted.
    *b. The patient's head was rotated.
    c. The patient's chin was elevated.
    d. The central ray was centered too superiorly.

23. An accurately positioned submentovertex (Basilar) cranial radiograph demonstrates:
    1. the foramen ovale and spinosum.
    2. the mandibular mentum anterior to the ethmoid sinuses.
    3. equal distance from the mandibular ramus to the lateral cranial cortex on either side.
    4. the mandibular ramus aligned with the long axis of the collimated field.

    *a. 1, 2, and 3 only
    b. 2 and 3 only
    c. 1 and 3 only
    d. 2 and 4 only

24. A patient is unable to sufficiently hyper flex the neck for a submentovertex (Basilar) cranial radiograph. How should the positioning setup be adjusted to obtain an optimal radiograph?
    *a. Align the central ray perpendicular to the IOML.
    b. Extend the patient's neck as far as possible and use a perpendicular central ray.
    c. Align the central ray perpendicular to the OML.
    d. The radiograph cannot be obtained.

25. Which of the following pertains to a submentovertex (Basilar) cranial radiograph taken with the vertex of the patient's head tilted toward the right side?
    a. The mandibular mentum will be turned toward the right side.
    b. The distance from the left mandibular ramus to the lateral cranial cortex is greater than the distance from the right ramus to the lateral cranial cortex.
    c. The mandibular mentum is demonstrated anterior to the ethmoid sinuses.
    *d. The distance from the right mandibular ramus to the lateral cranial cortex is greater than the distance from the left ramus to the lateral cranial cortex.

26. A poorly positioned submentovertex (Basilar) cranial radiograph demonstrates the mandibular mentum too far anterior to the ethmoid sinuses. How was the positioning setup mispositioned to obtain such a radiograph?
    1. The patient's neck was overextended.
    2. The IOML was not aligned parallel with the film.
    3. The central ray was angled too caudally.
    4. The patient's head was tilted toward the right side.

    *a. 1 and 2 only
    b. 1, 2, and 4 only
    c. 1, 2, and 3 only
    d. 4 only

27. A poorly positioned submentovertex (Basilar) cranial radiograph demonstrates the mandibular mentum posterior to the ethmoid sinuses. How could the positioning setup be adjusted to obtain an optimal radiograph?
    1. Tilt the patient's head toward the left side.
    2. Angle the central ray cephalically.
    3. Increase neck extension.
    4. Align the IOML parallel with the film.

    a. 1 only
    b. 2 and 4 only
    c. 3 and 4 only
    *d. 2, 3, and 4 only

28. For a parietoacanthial (Waters) sinus radiograph:
    1. the patient is positioned upright to demonstrate air-fluid levels within the maxillary sinuses.
    2. the MML is aligned perpendicular to the film.
    3. the central ray is centered to the acanthion.
    4. the OML is at a 37-degree angle with the central ray.

    a. 1 and 4 only
    b. 2 and 3 only
    c. 1, 2, and 3 only
    *d. 1, 2, 3, and 4

29. An accurately positioned parietoacanthial (Waters) sinus radiograph demonstrates:
    1. equal distances from the lateral orbital rim to the lateral cranial cortex on each side.
    2. the bony nasal septum in alignment with the short axis of the collimated field.
    3. the petrous ridges visualized inferior to the maxillary sinuses.
    4. the ethmoid sinus through the mouth cavity on an open-mouthed position.

    *a. 1 and 3 only
    b. 2 and 4 only
    c. 2 and 3 only
    d. 1, 3, and 4 only

30. An acanthioparietal (Waters) sinus radiograph can be distinguished from a parietoacanthial (Waters) sinus radiograph because:
    1. it demonstrates the bony nasal septum in alignment with the collimated field's longitudinal axis.
    2. it demonstrates greater orbital magnification.
    3. it demonstrates less distance from the lateral orbital rims to the lateral cranial cortices.
    4. the petrous ridges are visualized superior to the maxillary sinuses.

    a. 1, 2, and 3 only
    *b. 2 and 3 only
    c. 1 and 4 only
    d. 3 and 4 only

31. A patient is unable to sufficiently elevate the chin for a parietoacanthial (Waters) facial bone radiograph. How could the positioning setup be adjusted to obtain an optimal radiograph?
    a. Angle the central ray cephalically.
    b. The radiograph cannot be obtained.
    *c. Align the central ray parallel with the MML.
    d. Elevate the patient's chin as far as possible and use a perpendicular central ray.

32. A poorly positioned parietoacanthial (Waters) facial bone radiograph demonstrates the petrous ridges within the maxillary sinuses. How was the positioning setup mispositioned to obtain such a radiograph?
    1. The MML was not aligned perpendicular to the film.
    2. The patient's hand was rotated toward the left side.
    3. The patient's chin was tucked more than needed.
    4. The central ray was angled too cephalically.

    a. 1 and 3 only
    b. 2 only
    c. 3 and 4 only
    *d. 1, 3, and 4 only

33. A poorly positioned acanthioparietal (Waters) facial bone radiograph demonstrates the petrous ridges too far inferior to the maxillary sinuses. How could the positioning setup be adjusted to obtain an optimal radiograph?
    1. Depress the patient's chin.
    2. Align the central ray parallel with the MML.
    3. Align the MML perpendicular to the tabletop.
    4. Adjust the central ray angulation caudally.

    a. 1 and 3 only
    b. 2 and 3 only
    c. 1 and 4 only
    *d. 1, 2, 3, and 4